Constructing a Relational Cosmology

Princeton Theological Monograph Series

K. C. Hanson, Series Editor

Recent volumes in the series

David A. Ackerman
Lo, I Tell You a Mystery:
Cross, Resurrection, and Paraenesis in the Rhetoric of 1 Corinthians

John A. Vissers
The Neo-Orthodox Theology of W. W. Bryden

Sam Hamstra, editor
The Reformed Pastor by John Williamson Nevin

Byron C. Bangert
Consenting to God and Nature:
Toward a Theocentric, Naturalistic, Theological Ethics

Stephen Finlan and Vladimir Kharlamov, editors
Theōsis: Deification in Christian Theology

Richard Valantasis et al., editors
The Subjective Eye:
Essays in Culture, Religion, and Gender in Honor of Margaret R. Miles

Caryn Riswold
Coram Deo:
Human Life in the Vision of God

Philip L. Mayo
"Those Who Call Themselves Jews":
The Church and Judaism in the Apocalypse of John

Edward J. Newell
"Education Has Nothing to Do with Theology":
James Michael Lee's Social Science Religious Instruction

Mark A. Ellis, editor and translator
The Arminian Confession of 1621

Constructing a Relational Cosmology

edited by
Paul O. Ingram

Pickwick *Publications*
An imprint of *Wipf and Stock Publishers*
199 West 8th Avenue • Eugene OR 97401

CONSTRUCTING A RELATIONAL COSMOLOGY

Princeton Theological Monograph Series 62

ISBN: 1-59752-590-1

Cataloging-in-Publication data:

Constructing a relational cosmology / edited by Paul O. Ingram.

Eugene, Ore.: Pickwick Publications, 2006
Princeton Theological Monograph Series 62

vi + 124 p. ; 23 cm.

ISBN: 1-59752-590-1

1. Feminist theory. 2. Ecofeminism. 3. Cosmology. 4. Metaphysics. 5. Process philosophy. 6. Howell, Nancy R. I. Ingram, Paul O. II. Title. III. Series.

HQ1190 C667 2006

Manufactured in the U.S.A.

Contents

1 Introduction / 1
   —*Paul O. Ingram*

2 All is Not Lost:
   Solidarity and the Particularity of Love / 17
   —*Marit A. Trelstad*

3 Story, Forgiveness and Promise:
   Narrative Contributions to a Feminist Cosmology / 37
   —*Lisa Stenmark*

4 Dualism Without Domination:
   A Reinterpretation of Dualism for Ecofeminist Theory / 54
   —*Kathlyn A. Breazeale*

5 Existence is Relational:
   Contemplating Friendship With Nature / 69
   —*Stephanie Kaza*

6 Does Feminism Need Process?:
   Yes, No, Maybe, All of the Above / 91
   —*Marjorie Hewitt Suchocki*

7 Beyond *A Feminist Cosmology* / 104
   —*Nancy R. Howell*

Bibliography / 117

Contributors / 121

1

Introduction

Paul O. Ingram

Introduction

A T the Center for Process Studies' "Whitehead Summit" held in 2001 at the Claremont School of Theology, representatives of various process groups from around the world reflected on methods by which process thought might be made more visible and influential. One suggestion was that the process community should be more intentional about getting critical attention focused on books written from a process perspective, particularly those informed by, but not limited to, the process metaphysics of Alfred North Whitehead. Among the most important process books now in print deserving critical assessment is Nancy R. Howell's *A Feminist Cosmology: Ecology, Solidarity, and Metaphysics.*[1]

My admiration of Howell's scholarship and work with students began during her tenure at Pacific Lutheran University, where we were colleagues in the Department of Religion. During this time, she became, and still remains, my primary instructor in feminist and ecofeminist thought and practice. In her book, she writes as a Christian philosophical theologian whose work is informed by the ecofeminist insight that the domination of women and the domination of nature are interrelated injustices. Her work is also theoretically grounded in process philosophy, particularly the metaphysics and cosmology of Alfred North Whitehead, as well as the biological sciences. She addresses three important questions. (1) Can Whitehead's

---

[1] Nancy R. Howell, *A Feminist Cosmology: Ecology, Solidarity, and Metaphysics* (Amherst, N.Y.: Humanity, 2000).

process metaphysics make a contribution to feminist cosmology? (2) If so, what does Whitehead's process metaphysics have to offer feminist cosmology? (3) What does feminist thought, particularly ecofeminist thought, have to offer Whiteheadian cosmology? Her thesis—that Whitehead's cosmology is compatible with ecofeminist thought and offers a theoretical foundation for a constructive feminist theory of relations—is supported by three arguments.

First, Howell argues that feminist proposals of alternatives to patriarchal concepts of relations based on hierarchical assumptions may be strengthened by the comprehensiveness of Whitehead's philosophical perspective, because the metaphorical character of much feminist theory is intellectually unsatisfying and troublesome. The problem with metaphorical approaches is that they are vague and ambiguous and, perhaps, elitist when divorced from a theoretical framework that provides rational support for their meanings. But Whitehead was a metaphysician and cosmologist "as well as a poet who suggested metaphors congenial with feminist sensibilities and concerns."

Second, even though Howell thinks Whitehead's thought is compatible with ecofeminist thought, Whitehead's cosmology should not be appropriated uncritically. In agreement with Mary Daly's *Beyond God the Father*,[2] Howell asserts that Whitehead neither directly addressed sexism nor the institutional roots of sexism in patriarchal social and political institutions. Daley's warning is that feminists must be aware of imposing prefabricated philosophical constructs because women's thoughts and words must be their own. Howell interprets this to mean that feminists should not employ Whitehead's vision itself as a feminist philosophy, because "it is hardly feminist." Accordingly, while feminist theory may be informed by process philosophy, feminists should be conscious of its limitations so that women's concerns define the agenda of the conversation.

Finally, feminist thought and process philosophy are concerned with the universality of relationships in a profoundly inclusive sense. All things and events are mutually co-constituted by the universal web of interdependent interrelationships that course through the universe at every moment of space-time. As Stephanie Kaza argues in her contribution to this volume, Whitehead's theory of universal interdependent relationship is also supported by much Buddhist thought and experience, particularly the Buddhist doctrines of non-self (*anatman*) and "dependent co-arising"

---

[2] Mary Daley, *Beyond God the Father: Toward a Philosophy of Women's Liberation* (Boston: Beacon, 1973).

(*pratītya-samuptpāda*). For process philosophy and theology, the web of universal relationships includes God's interrelatedness with all things and events in the universe at every moment of existence. Therefore, a coherent theory of relations is a philosophical necessity for feminist thought in general, and ecofeminist thought in particular. Howell offers three arguments for this assessment.

First, an ecological perspective seems to have impressed itself upon contemporary scientific views of nature to the degree that it is difficult *not* to think of living things and events as interdependent and related. Given her training in the biological sciences, Howell well understands the technical details of the natural sciences that point to the interdependence of physical relationships in the universe. From quantum events to electrons, from atoms to molecules, from plants to animals, from human beings to God, to the architecture of the universe itself, no thing or event is discrete and disconnected from other things and events. A theory of relations is absolutely essential to any scholarship concerned with nature and the meaning of existence.

Second, since all feminist movements entail commitment to relationships, it is self-evident that feminist thought originates in discontent with patriarchal and hierarchal notions of relationships. Out of this discontent, feminist thought—in all its forms—has evolved into a constructive enterprise imagining new forms and styles of relating, a task that is crucial for the liberation of relationships and the emergence of women's selfhood (9-10).

Finally, as a Whiteheadian Christian philosophical theologian, Howell notes that the root metaphors of Christian faith are deeply relational. She agrees with Sallie McFague that the New Testament parables and the life of Jesus are expressions of the Kingdom of God in which persons are brought into relationship with one another and with God by God's grace.[3] That is, the historical Jesus reveals that it is grace that introduces love as the defining characteristic of relationships between human beings and between human beings and God. The root metaphor in Christian faith is relationship exemplified by love within the Kingdom of God. So even though not all feminist theory is Christian, Howell believes that Christian relational models support feminist "practice" (*praxis*) wherever it occurs.

Since Howell employs Whitehead's cosmology as a means of experimenting with her ecofeminist cosmology, chapter two of her book is a de-

---

[3] Sallie McFague, *Models of God: Theology for an Ecological, Nuclear Age* (Minneapolis: Fortress, 1987) 28.

scriptive analysis of a Whiteheadian feminist theory of relations. The goal of this chapter is to establish the compatibility of Whiteheadian philosophy and feminism and recommends Whitehead as a resource for feminist philosophy and theology. Chapters three, four, and five seek to advance her thesis by a consideration of human relationships with nature (Chapter Three), human-human relationships (Chapter Four), and the God-world relationship (Chapter Five). Each chapter discusses current contributors to feminist thought and follows with descriptions of contributions that Whiteheadian philosophy may add to feminist scholarship. The final chapters specify how Whiteheadian cosmology supports a feminist theory of relations.

Five Critical Responses

A Feminist Cosmology has elicited a number of critical responses that attest to the fact that Howell's ideas have engendered serious interest among feminists, both Whiteheadians and non-Whiteheadians, as well as historians of religions like me who practice theological reflection and find themselves in the Whiteheadian camp. My primary intention in this volume is to bring together five feminist voices into critical dialogue with Howell's work: three deeply influenced by Whiteheadian process philosophy (Marit Trelstad, Kathlyn Breazeale, and Majorie Suchocki); a Buddhist who spends much of her professional time reflecting and writing about ecological issues (Stephanie Kaza), and a feminist voice deeply informed by narrative theology (Lisa Stemark). To create space for this dialogue, I have tried to write this introductory section as descriptively as I can, since I do not wish to impose my voice on the conversation between Howell and her dialogical partners.

Of course, the mere fact of gathering a collection of essays and organizing them into an anthology does, in fact, impose some sort of interference on the conversation. Still, each essay is self-contained and focused, as is Howell's response to their conversation with her; and I have tried to stay out of their conversation and listen as carefully as I can. My listening has taught me something important about feminist scholarship that many male scholars too often miss when they encounter the pluralism of feminist thought. These essays bear witness to what I perceive as a "collaborative unity" within the "diversity" of feminist thought, which has the effect of structuring out academic egoism from scholarly discourse. By this I mean that there is a fundamental vision shared by these writers that is stretched and tested at the boundaries where new experiences and realities are encountered and diversities of theoretical interpretation are creatively advanced. In this "stretching and testing," there is no hint of competition between the contributors to this discussion. This seems to me to be characteristic of the best feminist philosophy and theology, as well as the best Whiteheadian process philosophy and theology, and is perhaps the most important source of the continuing vitality and influence of both modes of thought.

Marit Trelstad writes as a Lutheran systematic theologian grounded in Whiteheadian process philosophy who is deeply committed to feminist concerns and methodologies. Her essay, "All is Not Lost: Solidarity and the Particularity of Love," focuses on the tension between personal self-identity through time as a communal construction and the particularity of the communally constructed self's identity that she sees at work in Howell's feminist theory of relations. She agrees with Howell's appropriation of Whitehead's process metaphysics of internal relations and its applications to human relations with nature and with God. Furthermore, she agrees with Howell that building relations of solidarity between white feminist, Womanist, Mujerista, and Asian feminist theologians would be deeply enhanced by process philosophy's worldview. Yet, while Trelstad agrees with Howell's advocacy of building relations between diverse feminists by emphasizing the connections between different feminists communities, the question is how this is to be accomplished. It is here that Howell and Trelstad differ.

Trelstad argues that defining individuals within networks of community is important to feminists because: (1) a sense of boundary is important as a means of preventing women's entire identity from being completely subsumed by relations to others; (2) boundaries are important for addressing the reality of women whose personal agency is violated physically and mentally by others; and (3) boundaries are important for understanding and respecting differences and plurality, especially within networks of solidarity between Mujerista, Womanist, white feminist, and Asian feminists. By focusing on the role of individuals within relationships of solidarity, one apprehends how issues of particularity and loss add depth to understanding solidarity between women.

To make her case, Trelstad notes that the experience of particularity and loss are important concepts for a process discussion of solidarity. By "particularity," she means the unique network of characteristics and actions that constitute a specific moment of any actual occasion of experience, including the society of actual occasions that constitute individual persons or selves. But in pointing to the particularity of any event or actual occasion, one is simultaneously faced with the experience of loss. Moments of experience cannot be kept; no experience can be repeated. Loved ones die along with their unique ability to continue contributing to community. Whitehead called this aspect of experience "perpetual perishing." However, in contrast to Whitehead's emphasis, in their focus on the self as a communal construction, very few process philosophers and theologians and very few feminist writers have stressed the importance of individual moments or the loss of individuals in their theories of relations.

Trelstad regards this as an oversight for two reasons. First, defining women in terms of agency replaces static boundaries with a definition of an individuality based on creative self agency. The practical importance of this is that it provides better means of understanding why the physical violation of women's bodies is experienced in terms of loss of agency. It is better, therefore, for women to define themselves in terms of their abilities as creators and shapers of themselves and their world, particularly since the agency of women has been either ignored or denied as a contributing factor to the major accomplishments of humanity.

Second, even while all human beings are internally related, others cannot make decisions for us, and our own creative ability can never be entirely negated. In process metaphysics, the responsibility for decision and action is ultimately based on the agency of individuals. These decisions and actions "accumulate for good or ill" and are inherited in the present as a whole social reality that is built bit by bit on the foundations of past subjective decisions. What this implies for human-human, human-nature, and human-divine relationships is that an individual, always in relation and constituted by these relationships, is responsible for integrating her or his own positive or negative response to these influences. From a feminist perspective, this implies that even within systems of patriarchal oppression, voices of creativity and resistance can still manifest themselves, since the actions of others that influence us are never the last word.

Lisa Stenmark engages Howell and Whiteheadian philosophy from the standpoint of narrative theology in "Story, Forgiveness, and Promise: Narrative Contributions to a Feminist Cosmology." She agrees with Howell's thesis that process and feminist thought could mutually benefit from the insights of the other, but her essay expands on Howell's project by adding two narrative threads. First, she draws upon Paul Ricoeur's, Stephen Crites', and Stanley Hauerwas' analyses of narrative as it arises from the experience of time as a continually perishing present in relation to the past and future. The past is past, but continues through memory; the future has not happened, yet it exists through present anticipation;

the present is fleeting to the point of non-existence, but is held by our attention to it. Stenmark argues that this aspect of narrative is similar, but not identical, to Whitehead's notion of "perpetual perishing," because it is through narrative and narrative alone that human beings configure their lives.

Next, Stenmark suggests that narrative theology can be included in Howell's Whiteheadian ecofeminist theory because of what narrative can contribute to a feminist theory of community and practice. For example, narratives help resolve the tension between selfhood as particular and selfhood as a communal construction that she and Trelstad apprehend in Howell's work. In narrative one looks through the eyes of particular, often marginalized, persons while simultaneously including wide ranges of perspectives that entail the rejection of universal truth claims. The tension between the individual and the community is resolved because a single narrative can look at the world from multiple perspectives, combine them into a whole, without privileging any single narrative or even resolving conflicts between narratives. However, while not all narratives are equal, narratives make it possible to frame a story in a certain way even as it includes multiple perspectives—a process similar to what Howell calls "importance," meaning the organization of data in a particular pattern.

The question is, by what criteria can narratives be judged? Clearly, Stenmark notes, we need standards by which to evaluate our stories and the stories of others. The problem is that in a post-modern world it is difficult, if not impossible, to defend abstract, universal, independent criteria for deciding between different truth claims. To Stenmark, Howell's "modernist" suggestion that "the unity of the universe" can provide such a criteria is in need of post-modern correction, because Howell's criteria are very difficult to support unless one is already committed to process metaphysics or feminist principles. Furthermore, like all commitments, Howell's commitment to "the unity of the universe" is rooted in a particular (Whiteheadian) narrative. But in a post-modern age, commitment to cultural narrative is the opposite of the modernist assumption that truth lies in abstract principles.

Consequently, Stenmark concludes that the only way to evaluate narratives is by larger stories that serve as cultural narrative: a community's history, evolutionary theory, contemporary scientific cosmology, a particular religious tradition. Thus speaking as a "Christian" while others speak as Muslims, Hindus, Buddhists, or atheists opens up space for conversation. But abstract criteria for judging truth claims in principle are exclusive, because anyone who disagrees with my claims is "unreasonable." A post-modern approach is far better, Stenmark concludes, because when disagreements are based on a particular tradition they are based on different stories. Yet she argues that the relativism of post-modernism is not debilitating because a narrative approach opens up the possibility for understanding real differences and why they exist..

In "Dualism Without Domination: A Reinterpretation of Dualism for Ecofeminist Theory," **Kathlyn A. Breazeale** agrees with Howell's charge that metaphysical dualism represents an oversimplification of reality. But she disagrees with Howell's claim that dualism based on notions of superior/inferior is the sole source of oppressive hierarchies such as male/female, human nature/nature, and God/humanity that dominate contemporary culture. Breazeale's thesis is that some forms of dualistic thinking can provide positive models for conceptualizing relationships between entities that are distinct, but nevertheless not separate, that relate to each other without the domination of one over the other. She draws on three sources in support of her thesis: her feminist analysis of Hildegard of Bingin's theory of sex complementarity, Nahua (pre-Columbian Mesoamerican) concepts of duality, and Whitehead's conception of the relationship between body and soul.

Breazeale begins by noting that Hildegard's theory of "sex complementary dualism" was in sharp contrast to the theological anthropologies written before and after her work. Not only did Hildegard reject the traditional inequality between men and women, she developed her theory well before Aquinas quoted Aristotle's declaration that "the female is a misbegotten male." In her association of both men and women and body and soul with the elements of nature, Hildegard implied that human beings are separate from, yet interconnected with, nature. Thus while Hildegard retained the dualisms of man/women, culture/nature, and spirit/matter in her theology, these relationships were complementary for her rather than hierarchal. Accordingly, there exists no justification for the domination or exploitation of women and nature in Hildegard's theological anthropology.

Breazeale next describes the interconnection between the body and the universe in Nahua cosmology as so intimate that there is permanent reciprocal interaction between the universe and the body, which seems to her to be similar to Chinese notions of the continual balancing between *yin* and *yang* energies guiding natural processes and human beings to a point of harmonious balance between extremes. Thus both Nahua cosmology and Taoist cosmology teach that humans should live in harmony with natural forces by avoiding the extremes of one-sided dualism, which only disrupt the harmonies of nature and life in community.

Finally, Breazeale cites three aspects of Whitehead's concepts of body and soul and soul and world as a source for developing an ecofeminist theory of duality without domination. First, the body is not inferior to the soul in Whitehead's thought, because both body and soul are constituted by shared experiences in a relationship of reciprocity. It is this aspect of Whitehead's thought that undercuts the traditional association of soul/male and body/female dualities, because Whitehead does not assign gender to either body or soul. Accordingly, the primacy of the body in experience is foundational. Second, because the body is not inferior to the soul and women are not associated with the body, one cannot on the basis of Whiteheadian thought consider women to be inferior to men. Third, Whitehead's theory of how the soul and the world mutually constitute the other undercuts the hierarchy of humans over nature; since, in contrast to associations of the female body with nature, Whitehead argued that the human soul and the world are mutually interdependent. That is, the soul is distinct from the world, but not separate from the world.

In "Existence is Relational: Contemplating Friendship With Nature," **Stephanie Kaza** engages Howell as a Buddhist ecofeminist and environmental activist. As a Buddhist, she does not reflect on aspects of Howell's ecofeminist theory that address relationship with God. Instead, she concentrates on a selection of some of the issues Howell raises as a means of adding a Buddhist perspective to Howell's ecofeminism. Specifically, Kaza addresses three questions: (1) What is valuable about feminist cosmology? (2) Why value nature? And (3) What can Buddhism contribute answering these questions?

Kaza asserts that answering the first question is necessary for her to engage Howell's work as such. She agrees that Howell makes a strong case for feminist cosmology as an alternative to existing patriarchal frameworks, especially those that are based on concepts of God as male. She argues that the value of a feminist cosmology is that it creates opportunities for critiques of alienation, dualism, and hierarchy, which can also be supported from a Buddhist perspective. Consequently, both feminism and Buddhism are open to the sort of ecofeminist cosmology proposed by Howell, minus her Whiteheadian-Christian monotheism.

The second question is of great importance for Kaza: "Why value nature?" Of course, there are practical reasons to value nature—spiritually, recreationally, or ecologically—which are critical for slowing the accelerating destruction of the planet's forests, waterways, oceans, and atmosphere.

Here too, Kaza agrees with Howell's stress on relating human values to natural processes minus dualistic, hierarchal, and anthropocentric views that posit the superiority of human beings over nature. At the same time, valuing nature is simultaneously valuing human existence, given the fact that human beings are utterly dependent on, and interdependent with, the natural processes that constitute existence. Again, Kaza agrees with Howell: valuing nature is essential to the continued existence of human beings on this planet. So the primary focus on Kaza's essay is how Buddhist insights might be applied to Howell's ecofeminist cosmology as a means of highlighting the commonalities and differences between Buddhism's nontheistic world view and the Whiteheadian Christian monotheism supporting Howell's ecofeminist cosmology.

It is well known by those engaged in Buddhist-Christian dialogue that the Buddhist world view is relational, which Kaza describes in terms of the doctrine of "interdependent causation" or "co-origination" (*pratītya-samuptpāda*). This doctrine, accepted by all lineages of Buddhism, states that all phenomena exist interdependently, so much so that no phenomenon exists separately or independently. Because all things and events are mutually co-created by and implicated in all things and events, no thing or event is ever separated from other things and events. The corollary to this doctrine is that no thing or event can be said to have a substantial self remaining self-identical through time. This means that "self" as understood in most Western theology and philosophy is a problematic notion from a Buddhist perspective.

While Kaza acknowledges there are similarities and some overlap between the Whiteheadian ontology of becoming and Buddhism, she nevertheless believes Buddhists go further than either Whitehead or Howell. For Buddhists, the basic human problem is clinging to a false view of the self as separate and permanent. It is clinging to the illusion of separate selfhood that is the cause of "suffering" (*duḥkha*). Since the experience of suffering is caused by human beings seeking what is not part of the structure of existence, liberation from suffering which Buddhists call Awakening (*nirvāna*) is achieved only when human beings train themselves to see thorough the delusion of permanent selfhood: for if there is nothing to which to cling, the only alternative is to stop clinging to permanence of any sort. Furthermore, as the problem of suffering is a human generated problem, so too is liberation from suffering. Understood from this perspective, reliance on God as a source of liberation can only be another means of generating suffering. Since for Buddhists, nothing is permanent, there leaves little room for God, which contradicts traditional and non-Whiteheadian Christian notions of God as absolute and unchanging.

Kaza concludes by looking at Howell's notion of friendship with nature, and again notes similarities between Howell's cosmology and Buddhist understandings. Howell contends that friendship with nature has the capability of engendering liberation from false views of the self, which leads to empowerment of women's selfhood, from which friendship between human beings and human beings and nature emerges through a process of Be-Friending (as described by Mary Daley), which in turn reveals what Howell calls the "original woman," i.e., a hard earned self-created self forged through the process of breaking through oppressive views. According to Kaza, Howell's notion of "original woman" seems laden with ego construction which needs to be pared away for the even more liberated sense of the emptiness of "self," which Mahayana Buddhists call "Original Self."

Marjorie Hweitt Suchocki's essay is in the form of a question with four answers: "Does Feminism Need Process? Yes, No, Maybe, All of the Above." Citing Carter Heyward, Sallie Mac Fague, and Rosemary Ruether as feminist theologians who in varying degrees recognize process points of views in their work, Suchocki believes there is ample evidence in support of the Howell's conclusion that feminism needs process philosophy. In light of the apparent "yes" answer to her question, Suchocki's essay is an investigation of the "No, Maybe," and "All of the Above" of her essay's title, which is also her essay's thesis.

The "No" in Suchocki's title is an affirmation that feminism stands alone in spite of the relational assumptions of feminist analyses of the human condition. She points out that the feminist movement originated in the passionate concern of women to achieve full equality and dignity with men. This alone energized the work of early feminists like Charlotte Perkins Gilman, Jane Addams, Virginia Wolfe, as well as contemporary feminists like Simone de Beauvoir, Betty Friedan, Mary Daley, and Rosemary Ruether. To suggest that these women "needed" a metaphysics to support their work is in contradiction to the historical facts and the heroism exhibited in these women's lives. They recognized the interdependence of persons and the needs to change the structures of power, but this recognition did not require an analysis of the metaphysical basis of power in order for them to do their work. In the context of the history of the feminist movement, Suchocki argues that the suggestion that feminist theology needs process metaphysics to further its work seems to belittle feminist thought and practice.

But does process thought need feminism? Again, Suchocki's answer is "no," because, in itself, feminism adds nothing to Whiteheadian process thought. She points to evidence that even though Whitehead actively supported the women's rights movement of his time, their issues dealt mostly with women's right to vote. But feminism has moved considerably beyond Whitehead's time, and it would be wrong to suppose that he should have incorporated contemporary feminist insights into the development of *Process and Reality*. In fact, *Process and Reality* is not concerned primarily with social justice issues. Whitehead's question was, "What must existence be like at its core, given the internal relations and responsive nature of even tiny sub-atomic elements discovered by physics?" He wanted to rethink the nature of the universe in terms of the centrality of responsive internal relationships. Since Whitehead's system addresses the structure of existence that underlies all forms of organic and inorganic systems, changes in the social system are not relevant to the structure of his metaphysics.

Even so, Suchocki also argues that process thought might benefit from feminism. She points to the fact that Whitehead himself was not content simply to reflect on a cosmological scheme that could specify how things and events relate in the universe. In spite of the fact that Whitehead sought to give philosophical expressions to what physics and the biological sciences of his time were revealing about the physical structures of the universe, he also sought to apply his metaphysics beyond his training in mathematics and physics to education, religion, and social history. Since cosmology has its useful application to wider areas of concern, to every cosmology one should ask, "So what?" Since, by definition, cosmological theories are intended to be universally applicable, there are in fact no inherent reasons that Whiteheadian process cosmology and feminism should not have much offer the other.

Howell's suggestion is that if process thought is tested against the insights of feminism, process thinkers would discover and correct a tendency toward hierarchal thinking and value. Since hierarchal tendencies are not essential to the fundamental Whiteheadian analysis of relations, Howell concludes that feminist thought can help process thinkers rid themselves of all its modes of reflection that engender hierarchal systems and attitudes. However, Suchocki is not sure process thought gains anything from feminism in this regard. She points out that even ecofeminists slap mosquitoes and take antibiotics to keep dangerous viruses from infecting them. Furthermore, while feminists value diversity, feminists would not mind eradicating living organisms responsible for smallpox, AIDS, cholera, and tuberculosis. In terms of process thought, the justification for hierarchies is "intensity of experience," which follows from self-enjoyment at increasing levels of complexity. Whitehead believed there is no sharp line separating forms of life, since all events are interdependent and interconnected, which means that while all living organisms are intrinsically valuable in and for themselves, human life entails greater enjoyment, intensity, and complexity than is the case for non-human life, even though this is a difference in degree. So while boundaries between species are indeed blurred in process thought, and all forms of life have both intrinsic value and instrumental value for others, there remains a clear basis for determining levels of hierarchal importance and value in Whitehead's metaphysics. It is this aspect of process thought that Howell describes as deeply needing a feminist corrective.

Howell's corrective focuses is on the importance of diversity, which she thinks guards against the tendency she sees in Whitehead's thought to make intrinsic experience the determiner of value and the source of hierarchy. She appeals to Whitehead's notion that God receives every aspect of experience into God's "consequent nature" so that it is integrated into God's everlasting becoming. The diversity of finite experience thus enriches divine experience through its continuous introduction of complexity and intensity. Howell sees this as an undermining of hierarchy.

While Suchocki does not quarrel with Howell's wish to undermine hierarchy, she criticizes Howell's appeal to the necessity of diversity even in God's experience as a repositioning of the very hierarchy Howell seeks to destabilize. Is not any appeal to God an appeal to the top of a hierarchal ladder? Suchocki thinks that Howell's biologically oriented analysis needs to be supplemented by social analysis: the necessary combination of intrinsic and instrumental values along with complexity and intensity of experience means that all hierarchies are contextual and multiple. Since there can be no single hierarchal ordering, problems emerge not from hierarchal structures *per se*, but from simplistic, one dimensional structures that fail to observe the ambiguities that emerge in a relational ordering of society. Accordingly, in disagreement with Howell, Suchocki does not think that process thought needs feminist critique regarding the problems of hierarchy. Instead, she argues that the chief benefit feminist thought can provide to process thought is to raise the consciousness of process thinkers by persuading them to examine the ways in which patriarchal assumptions can twist even the radical relationality of process into oppressive social hierarchies.

Suchocki also adds that "maybe" feminism benefits from process thought. A number of feminist writers—in this volume represented by

Howell, Trelstad, Breazeale, and Suchocki—have pulled aspects of process thought into their critiques of dualism and non-relational thinking as an important source of oppressive hierarchies. Since non-relational thinking has been the metaphysical foundation of patriarchy, feminists must find a counter-metaphysics to dislodge patriarchy's non-relational foundations. Of course, this happens through feminist attacks on dualistic thinking, but feminists have as yet not fully recognized that these dualisms, e.g., subject/object, mind/body, man/woman/ God/human, human/nature, are themselves dependent on a metaphysics that asserts the primacy of self-contained atomic substances. A substance metaphysics creates an either/or world worldview, so that feminists must address themselves not only to polar oppositions but to the substance metaphysics that support these oppositions. It is here that Whiteheadian process metaphysics makes its most important contribution to feminist thought.

The final answer to Suchocki's question is "All of the Above." On the one hand, "Yes," because feminism and process thought are relational ways of thinking and by definition value interdependence. Yet historically, "No," because each tradition developed in quite different social locations and each can stand independently of the other. However, because feminist thought and process thought have instrumental value that affects the other, maybe feminism does need process thought and process thought needs feminism. It may be that dialogical interaction between feminist and process thinkers may strengthen the work of both.

Howell's Response:
Beyond a Feminist Cosmology

In her response, Howell not only seriously engages these five critiques of "A Feminist Cosmology," she herself engages her own book critically. She notes that her book is "a product of its time" that was rooted in the groundbreaking feminist and ecofeminist models of the 1980s and 1990s and that her ideas were circumscribed by the contexts and limits of these theological developments, just as her present work is contextualized by present developments in Whiteheadian, feminist, and scientific thought in interdependent relation with the feminist and ecofeminist models of 80s and 90s. Feminist and ecofeminist thought has undergone continual processive change, and thereby remained a highly creative and pluralistic philosophical-theological perspective. Howell embraces this pluralism as a source of inspiration for her own evolving ecofeminist thought, a pluralism that includes the five essays included in this book, as a means of moving her scholarship forward as she seeks accountability for greater clarification of ideas that she affirms contribute to establishing justice for women, nature, and "others at the margins." To this end, she specifies three proposals.

Her first proposal, drawn from the work of Sallie McFague's *The Body of God: An Ecological Theology*,[4] appropriates McFague's notion of "attention epistemology." Attention epistemology addresses the concern for particularity raised in Trelstad's essay and allows Howell to address issues of narrative in her evolving thought raised by Stenmark's essay. This is because attention epistemology requires refocusing the main attention of scholarship and observation from "self-interest" and "other-utility" to another creature, human or non-human, who possesses particularity and intrinsic value apart from its usefulness to the observer. Such a procedure intentionally moves the observer out of the perspective, presuppositions, and biases he or she brings to the observation of other creatures.

Howell cites Jane Goodall's work with chimpanzees as an illustration of attention epistemology at work in the natural sciences. Goodall violated the conventions of ethology by naming animals instead of numbering subjects, by referring to chimpanzees with personal pronouns and relative pronouns usually reserved for human beings, by describing the personalities of individual animals, by asserting that chimpanzees are rational animals, by interpreting their behavior in terms of emotions, and by positing purposes for their behavior. For Goodall, the relation between observer and chimpanzee entered human experience and becomes part of the human story, creating new insight that requires an adequate ontology, such as Whitehead's, coupled with a corresponding liberative *praxis*. In this way, the "chimpanzee narrative" suggests that community may be realized between human and nonhuman animals, which means that humans should extend just and liberating relationships with each other and with animals since human and nonhuman animals are interdependent.

Howell's second proposal is that ecofeminists and feminists need to establish a plurality of coalitions that give insight and perspective in the struggle for justice. She notes that when she wrote *A Feminist Cosmology* she sought to develop an argument for the rationality of using Whiteheadian metaphysics for developing a relational feminism, work already begun by Marjorie Suchocki, Catherine Keller, and Rita Nakashima Brock. Even so, and in agreement with Suchocki, when Howell wrote *A Feminist Cosmology*, she was—and still is—unwilling to argue that feminism must

---

[4] McFague, *The Body of God: An Ecological Theology* (Minneapolis: Fortress, 1993).

always be Whiteheadian. Nor is all Whiteheadian thought likely to be feminist. Furthermore, feminist scholarship must engage the pluralism of feminist movements—womanist, Asian feminist, and *mujerista* theology—as a means of building coalitions in the struggle for justice for all women, as well as a means of testing one's own particular ideas in dialogue with women of color. Coalition building therefore requires engaging in multireligious and multicultural dialogues. Howell cites the importance of her participation in Buddhist-Christian dialogue and her reflection about ecofeminism and chimpanzees in the developing processes of her own ecofeminist thought as examples of her own practice of coalition building.

Howell's third proposal calls for the refinement of feminist reflection on the nature and function of hierarchy and dualism by appropriating empirical evidence from the natural sciences to provide nuance. In agreement with Breazeale and Trelstad, Howell thinks that feminists need to explore in more systematic detail how "dominance hierarchy" is defined and how the science-religion dialogue, interreligious dialogue, feminist, and gender biased approaches may actually be using the same terms to refer to different phenomenon and ideas. There is much controversy in this aspect of ecofeminist thought and great need for clarification.

Conclusion

The critical essays collected in this volume are evidence that creative scholarship is a process that leads to creative transformation, In Howell's case—but also exemplified in the scholarship of the six essayists—creative transformation involves three interdependent ongoing collaborative projects: (1) exploring the potential of attention epistemology and narrative as methodological approaches to both whole and parts in nature as a means of balancing generalizations about ecosystems with insight from the depths of species and the individual lives of human and nonhuman creatures in nature; (2) seeking coalitions that give insight and perspective in support of the justice each of these writers in their distinctive ways advocate; and (3) refining reflection on hierarchy and dualism using empirical evidence from the natural sciences to provide nuance. I, for one, eagerly await more insight from her and her colleagues on these projects.

2

All Is not Lost:
Solidarity and the Particularity of Love

Marit A. Trelstad

YEARS ago, two pioneering women of bluegrass, Hazel Dickens and Alice Gerrard, had a reunion tour.[1] At one of their performances on Mother's Day, Hazel told a story of leaving her coal-mining hometown in West Virginia as a teenager. There simply was not work there and she had to seek employment in the nearby larger city of Baltimore. She recounted a moment, the moment of releasing her mother's hand, when she turned away from her home to leave. In her free hand, she had a lunch her mother had made and an extra pair of shoes. She said that she left confident in young naiveté, believing that all loves were equivalent and that she would find loves that were akin to that of home. Now, as an older woman, she looked back at that leaving moment. She had learned that each love was particular and completely unique. Hazel stated that if she had known then that she would never again find a mother's love, she would not have so easily let go of her mother's hand.

> The world is thus faced by the paradox that, at least in its higher
> actualities, it craves for novelty and yet it is haunted by the terror
> at the loss of the past, with its familiarities and its loved ones. . . .
> The ultimate evil in the temporal world is deeper than any specific

---

[1] Dickens and Gerrard's duo had the greatest success in the early sixties and it had been decades since they had performed regularly together. The concert to which I am referring took place on Mother's Day, May 11, 1997, at the Riverside Folk Festival, Riverside, California.

17

evil. It lies in the fact that the past fades, that time is a "perpetual perishing."[2]

In Whitehead's cosmology, death and loss is as pervasive as life as each minute, each moment and occasion, slips into the past, losing the ability to feel. The particularity and subjectivity of the moment is gone as it ceases in its own activity and becomes fuel for future moments. It is the grandest recycling project conceived. Each subject, or actual occasion, feels all past occasions, comes into being, and then "perishes" as an acting subject, offering what it has become as an object for future occasions. In one sense, a process model ensures some form of life after death, in terms of individual moments or events in time, since past events are literally incorporated into present moments. In another sense, the process model describes a continual experience of loss that is part of being participants in time.

In her book *A Feminist Cosmology: Ecology, Solidarity and Metaphysics*, Nancy Howell thoroughly describes Whitehead's process metaphysics of internal relations and its application to human relations with nature, with God, and with each other.[3] One emphasis within this more general scope focuses on building relations of solidarity between white feminist, Womanist, Mujerista and Asian feminist theologians. From a process world-view, all events and people are interrelated and thus there is a basis for asserting a form of unity between diverse women. Howell states that female friendships, in particular, have the power to unseat several patriarchal structures, and thus she advocates building more relations, emphasizing our connections to one another. She writes,

> The introduction of female friendship into a largely hetero-relational world makes visible the complexity of the world. Gyn/affectionate power makes women and men uncomfortable, because it shatters an oversimplified view of reality and forces us to reconsider our hetero-relational perspective. Female friendship inspires the process of personal self-formation and reintegration in light of the introduction of a Gyn/affectionate anomaly into our experience.[4]

---

[2] Alfred North Whitehead, *Process and Reality* (New York: Free Press, 1978) 340.

[3] Nancy R. Howell, *A Feminist Cosmology: Ecology, Solidarity and Metaphysics* (Amherst, N.Y.: Humanity, 2000).

[4] Ibid., 94.

Additionally, she writes, "When female friendships entertain diverse women's experiences as contrasts, creative and constructive alternatives for social, political, economic, and ecological relations flower."[5]

And yet, as Howell clarifies, friendship and real relations require distinct individuals, where the self is not lost in the relationship. One needs difference, individuals, and diversity in order to have a true dialogical relationship between people. Diverse voices together bring contrast; they spark novelty and new possibilities, because they allow unresolved tensions in the conversation to remain. Within process thought, the presence of distinct individuals in relationship is an opportunity for richness of experience and new forms or patterns of existence. She writes, "Feeling complexity as contrasts, instead of dismissing (or negatively prehending) complexities as anomalies or incompatibilities, intensifies experience."[6]

In reference to what process philosophy offers to the discussion of solidarity and distinction between diverse women theologians, Howell argues that one must maintain an integral balance between the individual and the community in which the self is created. One is not dissolved in relations nor overly differentiated and alone. Both distinction and interconnection are necessary in solidarity. Howell's voice is a welcome addition to this particular conversation within process thought. Thus far, process writers have primarily focused on the relational aspect of existence to the detriment of a clear understanding of distinct, agential individuality. Because it has engaged philosophies and theologies that do not understand the world as a system of relations, process thought has rightly underscored the web-like relational matrix that sustains all of life. Within the context of most Feminist, Mujerista, Asian feminist and Womanist thinkers, however, the intense relationality of life is an assumed foundation of thought, and thus the definition of the individual in relation needs further clarity. Thus far, many of us have asserted the value or necessity of maintaining a sense of self within relationships, but few have given clear terms by which one may define the self without creating an oppositional "other." Although Whitehead cannot give every answer for a feminist cosmology, as Howell acknowledges, his work provides metaphysical precision of thought regarding self-definition within a thoroughly relational system.

But why look to defining individuals within these networks of community? First, a sense of boundary or distinction has been important to many feminist writers in order to prevent women's entire identity from

---

[5] Ibid., 96

[6] Ibid., 94.

being completely subsumed by relations to others. Catherine Keller and Howell use the term "soluble" to describe women dissolved within their relations. Howell writes, "The soluble self so immerses herself in relationships that her selfhood literally dissolves in the work of dependence and sustenance of the male separate self."[7] A sense of boundaries, individuals, or "self" seeks to prevent defining women only in terms of their value to others. Second, boundaries are important for addressing the reality of women whose personal agency is violated mentally and physically by others in overt and covert ways. Lastly, boundaries are important for understanding and respecting difference, especially within networks of solidarity between Mujerista, Womanist, white feminist and Asian feminist theologians. Even as we form a coalition of women theologians, we must maintain and stress the distinctions and gifts each person and each group of theologians bring to the table. Throughout her book, Howell presents, without solving or dissolving, the distinct voices within this dialog. From a process and feminist framework, however, this discussion of boundaries is not about drawing arbitrary or false lines of distinction between people. It is, rather, centered on a discussion of creative personal agency and its basis.

One of the most important caveats that Howell raises, in terms of this discussion, is that white Feminists have often concentrated on individual forms of liberation and philosophical notions of the "self" whereas Womanist writers have stated that they have always sought the liberation of the entire community from racism, patriarchy, and classism.[8] Thus, Womanists have claimed that debates over philosophical definitions of "self" have been white women's questions. Perhaps it is due to being a white feminist, but I find that these two discussions need not be polarized. I would concur with Howell that the distinctiveness of each individual within a relational context builds the sense of community and solidarity. The health of one depends on the health of the other. Howell evenhandedly addresses the value of both community and the self. Throughout this essay, however, I will choose to focus on the value of individuals within the community. I will establish a process definition of the self as emerging from its relational context through subjective agency. By focusing on the role of individuals within relationships of solidarity, one begins to see how the issues of particularity and loss add depth to understanding solidarity between women.

[7] Ibid., 97.
[8] Ibid., 81, 85.

Whitehead's non-substantialist metaphysics describes reality in terms of *dialogical action* between individuals. Thus, one cannot draw static, false boundaries between people, and yet real distinction exists between other-agency and self-agency. There is reciprocity and internalization of other's actions, but this does not compromise the agency of the individual. Other philosophers and theologians, from the existentialist movement, have also defined the self in terms of agency. Examining these philosophies will bring further insight into how an agential definition of the self needs to be coupled with a philosophy of interconnectedness in order to advance solidarity. Defining the self in terms of action holds many possible solutions to Howell's question of how one remains in solidarity while honoring individual voices in the dialog. In a more general sense, this understanding locates and defines the individual in a way that strengthens and enriches community, rather than compromising it, by stressing realistic boundaries and personal responsibility for just actions.

Particularity and loss also serve as important concepts for a process discussion of solidarity. By "particularity" I mean the unique network of characteristics and actions that define a specific moment or person. Recognizing the particularity of any event or actual occasion, one is naturally faced with the issue of loss. These moments cannot be kept; there is a perniciousness within experience in that it cannot be repeated. Our loved ones die and their unique ability to continue contributing to the community and their own being is lost. For Whitehead, the concept of loss or the "perpetual perishing" of individual moments is monumental enough to warrant his final chapter on God. By contrast to Whitehead's emphasis, very few process thinkers and fewer feminist process writers have stressed the importance of individual moments or the loss of individuals within a relational scheme. This is a great oversight. Even as Howell addresses the need for distinct, non-soluble selves in relation, I would add that a specific discussion of *loss* is also necessary to understanding solidarity. Together, the concepts of particularity and loss may serve as additions to the strong basis for dialog Howell has encouraged between women theologians. An appreciation of the particular, individual uniqueness of each person leads to allowing true difference to exist simultaneously with togetherness, to welcoming the beauty and strength of contrast, and to an ethic of responsibility. Loss emphasizes the value of a particular moment or perspective and, knowing this, one can face the loss of individual lives and creativity with a sense of precious honesty.

Evil and Loss in Process

There is loss in the selection process of what will be useful and relevant to the present. More than once, Whitehead refers to this part of the process of human experience as "evil." "The evil in the world is that those elements . . . with individual weight, by their discord, impose upon vivid immediacy the obligation that it fade into the night."[9] There is another form of loss in the process as well. The present internalizes the past, but past events are not active in the present as though they are committee members deciding the fate of the present.[10] The past has ceased to be subjective and active and has become objective, unable to act and create any longer. To describe the present moment's acquisition of this past, which has become "dead facts," Whitehead poetically paraphrases Psalm 127:2, "He giveth his beloved sleep."[11] So all is lost.

Or is it? For Whitehead, religion is the avenue by which humanity seeks to address loss. He writes, "The most general formulation of the religious problem is the question whether the process of the temporal world passes into the formulation of other actualities, bound together in an order in which novelty does not mean loss."[12] This statement provides an introduction to his last chapter of final interpretation entitled "God and the World." The "consequent nature" of God is the part of God that feels all that has become in the world. While temporal events or beings can incorporate only some of the past, God does not have to solve discord and therefore can include all the past into God's being. In other words, God has an infinite ability to contain contrast within God's being. What this means, in regard to loss, is that nothing is lost within the life of God. All moments are felt perfectly by God, with no negation of intensity or quality, and thus are preserved forever within God's being.[13] God assures

---

[9] *Process and Reality,* 341.

[10] Whitehead writes, "The present fact has not the past fact with it in any full immediacy. . . . In the temporal world, it is the empirical fact that process entails loss: the past is present under an abstraction"; ibid., 340.

[11] Ibid.

[12] Ibid.

[13] Marjorie Hewitt Suchocki has described a process conception of the afterlife based on this idea. She argues Whitehead's point that God can contain all contrast within God's self. When God feels the past moments, God is able to incorporate them perfectly within the life of God. Thus there is a basis for a Christian understanding of the afterlife. Explaining this further, she includes a chapter entitled "Subjective Immortality." See Marjorie Suchocki, *The End of Evil: Process Eschatology in Historical Context* (1988; reprinted, Eugene, Ore.: Wipf & Stock, 2005) 152. In writing a Christian, process theology of the af-

humanity that all is not lost. God saves the shipwreck of events from the shores of time.[14]

Howell's writing concerning relationality and solidarity emphasizes divine, natural, and human carriers of present experience into the future. All the past is integrally linked to the becoming of the present. Through internal relations with others, one experiences contrast, value, importance, creative possibilities, and more. This relatedness, to God and to the world, is a key aspect of process and feminist thought, which Howell picks up in her chapter titles, "Relating to Nature," "Relating to Each Other," and "Relating to God." Even perishing or an afterlife may be seen from a relational perspective. In reference to how humans relate to God, Howell recounts that some ecofeminists may argue that death is simply a final return to the greater network of life from which we emerged. Rosemary Radford Ruether proposes that one need not fear death since it is the time when one rejoins the great Matrix of life from which we have emerged.

> As we gaze into the void of our future extinguished self and dissolving substance, we encounter there the wellspring of life and creativity from which all things have sprung and into which they return, only to well up again in new forms. . . . The small selves and the Great Self are finally one, for as She bodies forth in us, all the beings respond in the bodying forth of their diverse creative work that makes the world.[15]

For much of Christian theology, which has focused exclusively on the salvation of individual souls to the detriment of understanding humanity's connection to the salvation of all creation, this collective vision of salvation and regeneration is a needed corrective. It takes into account a more holistic vision of a new heaven and new earth envisioned in many New Testament accounts. This vision of the afterlife, however, does not take into account the real experience of loss, of sorrow in this life at the loss of singularity. When someone dies, her or his perspective is gone forever. Their influence is incorporated into the lives of others and the life of God, and thus has an arguable continued existence, but that particular person or event has ceased its own subjective creativity.

---

terlife, she claims that the past moments may continue to have some sort of life themselves within the life of God.

[14] This is Whitehead's metaphor from the beginning of "Part V: Final Interpretation," in *Process and Reality*.

[15] Rosemary Radford Ruether as quoted by Howell, *A Feminist Cosmology*, 124.

I can remember my grandmother looking across the lake, counting the boats. I can recount the descriptions she gave the scene, her gestures and body position, how she liked to cross her arms and hold her elbows in her palms. I shared the overall context we were in: a hot evening, the sound of wind in the gnarled oaks, the smell of green bean casserole, a history with this lake cabin where we both had grown from babies to adults, albeit seventy years apart. I even have "Trelstad eyes," inherited genetically through her. And yet, I could never see directly through her eyes and experience, creating the scene exactly as she would. Now that she has died, that perspective, the event of her self, passes further and further into less clarity. While relationality is the vehicle through which the past is incorporated into the present, process thought also explains the fading of the past's influence as a particular moment or event becomes less proximal. The influence is always there to some degree, but it has less impact and clarity as time continues.

While the self is constituted by its relational context, Whitehead is clear that past actual entities do not have agency in the present moment. They have already "perished" in terms of their subjective agency and, in the present, are objects for the present moment's process of becoming. The subjective aim, or the present moment's action towards its own becoming, is the shaping and decisive agent until it too comes to completion or "satisfaction," in Whitehead's terms. Additionally, he writes that one inherits the past, but one prehends the past from one's own perspective, based in the present. The past cannot be entirely repeated or re-experienced. Whitehead thus describes a past that fades even while he stresses the importance of relationality between the past, present and future.

Whitehead's key contribution to philosophical discussions is his description of reality in non-substantialist terms. Instead of tiny bits of matter relating externally to each other, the world is composed of energy events that are interrelated. Within our century, physics has also described all physical matter in terms of energy. As stated earlier, Howell seeks a description of the self and other in solidarity such that the individual is neither consumed nor overly separate from her community context. From a process perspective there is the additional question that needs to be addressed in order to discuss particularity amid solidarity: what sense of boundaries does one have if the world is made of interconnected energy events?

Reality in Action:
Whitehead and Existentialism

Like Howell and Whitehead, I am pursuing a mechanism whereby one could claim to be both interconnected and distinct. The two integrated, key ideas I glean from Whitehead's work for the purposes of discussing the solidarity of distinct individuals are that reality is composed of action and all of reality is interconnected. Without the first aspect, one loses the distinctness of individuals. Without the second, individuals are seen as opponents struggling to dominate each other.

Whitehead offers the view that reality is composed of a rhythm of actions. Action is so critical to his understanding of reality that he calls all moments of time "actual occasions" or "actual events." *It is the subjective action of each event or moment that defines its existence.* The boundary of an event or being cannot be described with complete accuracy in terms of a physical separation. Whitehead addresses this in his book *Modes of Thought*. He writes that

> if we are fussily exact, we cannot define where a body begins and where external nature ends. . . . Consider one definite molecule. It is part of nature. It has moved about for millions of years. Perhaps it started from a distant nebula. It enters the body; it may be as a factor in some edible vegetable; or it passes into the lungs as a part of the air. At what exact point as it enters the mouth, or as it is absorbed through the skin, is it part of the body? At what exact moment, later on, does it cease to be part of the body? Exactness is out of the question. It can only be obtained by some trivial convention.[16]

So, physical or substantial boundaries cannot be considered definite. Additionally, in a process framework, all events are internally related, constituting one another. In light of this, how can one even begin to discuss individuality at all within a process? All actual occasions come to being in the context of all preceding moments of time and then offer what they have become to all future events. While past moments had acted subjectively in their own creation, they then reached "satisfaction" or completion, wherein they became objects for future occasions. The key point here is that while all events are internally related to one another, a past event is not *subjectively* active in the present moment. As stated earlier, the past is not within present moments like an active, deciding committee. Boundaries in this kind of system are created through each moment's action toward its

---

[16] Whitehead, *Modes of Thought* (1938; reprinted, New York: Free Press, 1968) 21.

own becoming, even as it is incorporating and affecting its entire relational context. This may be best illustrated through a metaphor.

I may consult my friend Laura regarding a decision. In this conversation, I will be influenced through her ideas, physical reactions, this history of our past discussions, the weight her opinion has for me because of the respect I have developed for her, and my confidence that she has my best interest in mind. If we are sitting together, we will also be sharing physical connections: the air we are breathing, our proximity, and the influence of our entire physical context. In all these ways, Laura is highly influential in my decisions, and all these elements factor into the process. In many ways, conscious and unconscious, she is influencing my decision. Nonetheless, Laura cannot be inside of my being, creating my decision for me. I absorb our interaction, come to some decision, and offer it to both of our futures. This is distinction in action. Each moment in time, or actual occasion, has this subjective nature to it, and, in this agential sense, it achieves its own being. While words such as "subjective" and "achievement" may typically connote conscious choice, this is not the case in Whitehead's system. Every element in the world participates in these energy events; some simply have a greater capacity to incorporate more contrast and novelty.

Lest it sound as though process thought also falls into the trap of asserting a separative, patriarchal sense of self definition,[17] one must keep in mind that process discussions of self have utilized two different foci. One discussion of the self begins in the model of the individual actual occasion and then generalizes an understanding of individuality based on the most elemental aspect of the process model. A second process concept of selfhood begins with an analysis of the self as a conglomerate of countless actual occasions occurring simultaneously, receptive to the proximal influence of all the other past moments. In this way, one can best understand how the network of the body and mind is thoroughly interrelated.

This second basis of understanding the self is primarily used by Catherine Keller, in her book *From a Broken Web*, when she writes that the "spider self" understands "the multiplicity of self-occasions." She quotes Robin Morgan, "speaking for the archetypal spider-self: 'Let me sit at the

---

[17] The use of the word "separative" in relation to patriarchy comes from Catherine Keller's use of this term in her book, *From a Broken Web: Separation, Sexism and Self* (Boston: Beacon, 1986), where she claims that nothing is truly separate in an interconnected world and so the attempt, by men historically, to define themselves as lone heroes has led to domination and suppression of women. She writes, "In contrast to the soluble self, which dissolves in relation, the separative self makes itself the absolute in that it absolves itself from relation. It brooks no other subjects and so it turns them into the nonsubjective other, the object, whenever it can" (26).

center of myself and see with all my eyes.'"[18] In this section of her writing, Keller is discussing the construction of the human being as a composite of millions of energy events, like the cells of our bodies. The total grouping of these events would be the multiple self. Even within this second definition of self as conglomerate, however, most process writing on the self, including that of Whitehead in *Process and Reality*, suggests that in higher order conglomerates there is some organizing factor. In other words, while a human being is a vast number of actual occasions occurring simultaneously, there will be an organizing actual occasion coordinating the various parts and systems. This is unlike a "lower order" of conglomerate, such as a rock, which has no such coordinating principle. Thus, process thought may express the self as both an interrelated group of events and/or as a singular actual occasion; but in more ordered systems, such as a human, the self has some ultimate coordination, aptly described by generalizing a single actual occasion.

Keller and Howell both emphasize that a feminist process discussion of self in relation cannot condone a patriarchal "separative" self. There is no such thing as a "self-made man," to use a common phrase. All being comes from a wide web of relationality created by interdependent subjects. While the idea of being "self-made" is false in terms of Whitehead's concept of internal relations, it is not a totally inaccurate description of Whitehead's conception of agency as the foundation of existence. Again, his most basic philosophical claim is that "only actual occasions can act," and these extremely brief and interconnected moments in time are the basis on which all creation rests. And while the present is interconnected, in terms of receiving the past and giving to the future, only the present has agency to create. Even God cannot act as an agent in another's moment of becoming. God can only offer persuasive influences to the present and respond to the world's decisions. Thus, in process thought, one does indeed have self *made* moments, but the influences from which the self arises are all "imported" from the past or from God. In other words, one *can* argue for a self-made woman, but what she is made of is not her own. A person is a true unique and distinguished individual, separate in agency in the act of self creation, even while connected to the past and future and the being of God.

The philosophy of reality being composed of actions rather than substances or essences is not Whitehead's philosophical concept exclusively. The existentialist movement also connected human reality with continu-

---

[18] Ibid., 250.

ous actions toward self creation. Marjorie Grene writes, "This self-creation—the making of one's essence from mere existence—is demanded of each of us because, according to existentialism, there is no *single* essence of humanity to which we may logically turn as standard or model for making ourselves thus or so."[19] Jean-Paul Sartre obsessed over defining the self in terms of action. Catholic theologian and philosopher Gabriel Marcel, also rooted in the existentialist tradition, carefully mapped out how this definition of the self affected the self-other relation. Like Whitehead, Marcel proposed a sort of dialogical creativity between the self and other. Both Whitehead and these existentialists understand existence as creatively active at its core. They all define the self through action rather than physical, static, or artificial boundaries. Also interesting, for the purposes of this paper, is how this core understanding affects each philosopher's perspective on the relation between the self and community. Marcel and Whitehead couch their view of the individual within a philosophy of constructive interrelatedness, whereas Sartre generally views relatedness as a necessary evil against which one strives to emerge.

Sartre locates freedom in the necessity of our actions toward ourselves and toward others. We continuously create our world. Our free action, what we *do*, is all that we *are*, all that we *have*.[20] Existentialism's reliance on looking only to existence for meaning is illustrated well by the film *As Good as it Gets*.[21] While waiting for psychiatric counseling, one of the main characters looks at the other people in the waiting room. They are nervous, depressed, sad, anxious; some of their hands are shaking, or their legs are twitching. Over-loudly and somewhat sardonically, he interrogates the group, "What if this is as good as it gets?" There is no elusive "wholeness" or divine essence on which to rely—life simply is what you make it. Certainly there is a freedom that comes with a stress on human action; but, in Sartre, this freedom is heavily laden with responsibility and a certain desperation, since our selves are in a state of perpetual action without ever achieving short moments of resolution. While Whitehead affirms that life is composed of action, he also concludes that moments do come to some satisfaction, some completion, and thus this conveys a less desperate or frantic pace than Sartre. For Sartre, we strive and strive. "The

---

[19] Marjorie Grene, *Dreadful Freedom* (Chicago: University of Chicago Press, 1948) 41.

[20] Grene summarizes: "Freedom is total, yet rooted in a determinate, historical situation; dread in the face of such freedom; and the concealment of the dread in the comforting frauds of everyday existence—such is the nexus of ideas that make up the core of the existentialist's conception of human life"; ibid., 56.

[21] *As Good As It Gets*, directed by James L. Brooks (Columbia/Tristar Studios, 1997).

paradox is not that there are 'self-activated' existences but that there is no other kind. What is truly unthinkable is passive existence; that is, existence which perpetuates itself without having the force either to produce itself or preserve itself."[22]

One striking difference between process thought and Sartre emerges in how Sartre views the interconnectedness that he grudgingly and despairingly acknowledges. The self cannot know its existence without others, but this is seen as a constant source of angst. For Sartre, all relations to others' striving for existence begin with negation. The other is *not* me, and her or his activity threatens to engulf me as an object, canceling my efforts at subjectivity. "Between the Other and myself there is a nothingness of separation . . . as a primary absence of relation, it is originally the foundation of all relation between the Other and me."[23] Along with this external relation with the Other, one is profoundly affected, internally, by the being of others in two ways that Sartre proposes. First, everything one interacts with signifies the existence of another person's action. One's past and entire physical environment are signs of previous people's choices and actions on a small scale or of societal structures on a larger scale. One is surrounded by the results of other's actions. Second, one can only see oneself through the eyes of others, in a snapshot of one's being. Since one is always involved in the ceaseless activity of creating, one cannot obtain a static "picture" or view of oneself and, instead, must rely on other's impressions alone. For Sartre, the other's eyes are threatening because they alone can define and reflect one's being, making all of one's earnest striving into a caricature.

As a Christian existentialist, Gabriel Marcel offers several new perspectives on self and relationality while affirming, with Sartre, the importance of building philosophy from our existential experience. He does not see the individual-other relation in terms of battle, but rather in terms of creative hospitality to the other within one's being. Thomas Van Ewijk writes:

> 'L'enfer c'est les autres' (Hell is the other person), says Jean-Paul Sartre (*No Exit*). . . . Against this certainty of Sartre's, Gabriel Marcel puts his brand of certainty . . . Some sort of a meeting must be possible in which we discover in our fellowman [sic] something other than our hell. And the more we stress the personality of the 'I' in our

---

[22] Hazel E. Barnes, in her translator's preface to Jean-Paul Sartre, *Being and Nothingness: An Essay on Phenomenological Ontology* (London: Metheun, 1957) lvi.

[23] Sartre, ibid., 230.

being, and the special nature of our existence, the more it becomes clear that confronting the 'I' there is someone else, a 'thou,' for whom I must open myself, if I want to discover a little who I really am myself.[24]

For Marcel, our everyday experience reveals being, which finds its Absolute Being in God. Whereas Sartre entitles a book "Being and Nothingness," Marcel provides a book of philosophical reflections entitled "Being and Having." In our self-creation of our being and in participation with the beings around us, we do not experience an overwhelming nothingness, but rather, Marcel affirms, a plentitude of being. Only the isolated individual, ego-centric or self-conscious, is caught up in her/his own self, feeling threatened, whereas the healthier self can open up to fuller participation with others and, in this permeable feeling of self-with-others, find freedom.

When it comes to a definition of self, Marcel approaches this issue mainly in his discussion of action. Actually, his discussion of "person" and "action" are entirely interwoven several places in his writing, most notably the chapter entitled, "Observations on the notions of the act and the person" in the book *Creative Fidelity*. Like Sartre, Marcel holds that actions define or create the person and this is seen, in particular, through responsibility; an action is mine and I must claim it. He utilizes the example of telling a lie and how that commits one to acknowledging that I, as subject, committed this action. In this manner, action is not only a *doing* but "the essence of the act is to commit the agent. . . . what is characteristic of my act is that it can later be claimed by me as mine; at bottom, it is as though I signed a confession in advance . . . ."[25] A subject is defined by agency, by actions. It is here in Marcel's philosophy that an ethic of responsibility emerges, and this affirmation of both agency and responsibility will be raised later in this paper in relation to dialogue between women theologians.

Instead of Sartre's being-in-itself, Marcel proposes a being-in-society. He uses the words "participation," "permeability," "intersubjectivity" and *co-esse* to describe how we live together. Intersubjectivity is not a kind of communication between two radio transmission towers or posts but rather there is a mingling in the middle of two selves that takes on its own

---

[24] Thomas J. M. Van Ewijk, *Gabriel Marcel: An Introduction,* trans. Matthew J. van Velsen (Glen Rock, N.J.: Paulist, 1965) 65–66.

[25] Gabriel Marcel, *Creative Fidelity,* trans. Robert Rosthal (New York: Farrar, Strauss, 1964) 107.

being.[26] Even while he stresses this participation in the being of others, Marcel does not want to negate the self and its importance. Marcel raises an interesting image to illustrate what he means by having a certain permeability to the self: "What we are driving at is not a kind of porousness, like that, for instance, of a sponge. It would be better to think of that sort of aptness to be influenced, or that readiness to take impressions, which is called in English 'suggestibility' or 'impressionability . . .'"[27] Connected to this idea of receptivity is his notion of the self as one who feels at home enough in oneself to be hospitable, to open oneself to another. This ability to create a place for the self is yet another activity of the person as agent. Love, or the divine "Thou" underlying all mutual relations, becomes "the breaking of the tension between the self and the other, appears to me to be what one might call the essential ontological datum."[28] In Marcel's philosophy, God becomes the Absolute Thou whom we experience. The more we open ourselves to the mystery of being, the more we encounter the ultimate co-presence in our lives.

Mary Daly, whom Howell addresses extensively, also utilizes some aspects of existentialism in her book *Beyond God the Father*, where women learn to step out of noun-al, non-existence defined by patriarchy and to boldly act themselves into existence. She advocates moving "in the current of . . . the Verb" of *becoming*, rather than *being* defined by patriarchy.[29] For Daly, life and being are equivalent with action and agency. Throughout her book, she works with the concepts of fixed, death-dealing gender stereotypes that have been reified by positing a God who embodies and enforces patriarchal rules. In these caged definitions, women are denied the means and, hence the will, to fully live. They lose the power of their own creative being and naming. Other feminist theologians have also claimed that it is our human desire to objectify people and nature, through static, possessive definitions, that leads to justifying the domination of all we define. Instead, Daly advocates that what is truly divine is the activity of love, an activity of becoming, that resists categorization since it is perpetually changing: "This Verb—the Verb of Verbs—is intransitive. It need not be conceived as having an object that limits its dynamism."[30] Additionally,

---

[26] Marcel, *The Mystery of Being*, trans. G. S. Fraser (Chicago: Regnery, 1960) 1:124.

[27] Ibid., 1:178.

[28] Marcel, *Being and Having*, trans. Katherine Farrer (1949; reprinted, New York: Harper & Row, 1965) 167.

[29] *Beyond God the Father: Toward a Philosophy of Women's Liberation* (Boston: Beacon, 1973, 1985) 197.

[30] Ibid., 34.

she writes, "In hearing and naming ourselves out of the depths, women are naming *toward* God, which is what theology always should have been about. Unfortunately it tended to stop at fixing names *upon* God, which deafened us to our own potential for self-naming."[31] In these ways, of defining a new selfhood through verbal activity, Daly works closely with existentialist understandings of the self. As with Sartre, Daly's key opponent to existence is still nothingness. For Daly, patriarchy has denied women agency and, therefore, existence, and so it has forced them into nothingness. She writes, "The freedom-becoming-survival of our species will require a continual, communal striving in be-ing. This means forging the great chain of be-ing in sisterhood that can surround nonbeing, forcing it to shrink back into itself. The cost of failure is Nothing."[32]

Reviewing Sartre, however, one can see that static definitions alone are not the cause of dualistic and oppositional self-other thinking. He has a philosophy of self created in action, not by static definitions or essence, and yet he asserts a patriarchal notion of life as a constant battle of winning and losing, rather than one of interconnection. This "battle" mentality appears to be the key cause of his dualistic, self-against-other, patriarchal attitude. Thus, a sense of constructive solidarity between individuals requires more than only understanding the self as composed of action. One needs to also base this agential understanding of the self in an overall philosophy of interconnection. Marcel and Whitehead both do this and hence achieve a more useful, well-balanced conception of the self within a relational framework. Marcel gives better poetic, approachable description to the self-other relationship, whereas Whitehead gives metaphysical clarity to the precise interactions within this relationship. Daly also urges that women, emerging into their own being, should not follow Sartre's example and see the process of becoming as necessitating an enemy, an "other" to oppose. Non-being itself is the only enemy.[33] Sartre, Daly, Marcel, and Whitehead each affirm that others' actions have an enormous influence on the individual. The agencies of self and other are seen in dialog with

---

[31] Ibid., 33.

[32] Ibid., 198.

[33] In this book, written early in her career, Daly does not find surer footing in a philosophy of interconnectedness. It is important to note, however, she goes beyond an explicit conversation with existentialism and theology in her later works, seeing her efforts here in *Beyond God the Father* as still too deeply rooted in patriarchal categories of thought. In her later writings, she maintains her urging of women into the process of becoming; but a philosophy of interconnectedness emerges in a pronounced manner in such works as *Gyn/Ecology: The Metaethics of Radical Feminism* (Boston: Beacon, 1978).

one another, or in fierce competition, if one is describing Sartre. What Whitehead's philosophy offers is a co-existent philosophy of interrelatedness to explain why there is no true "other." In order to cure our sense of encounter as battle rather than solidarity, one must see life as based in action, rather than fixed categories, and also incorporate a sense of the interconnectedness of all of life.

Agency, Particularity, and Loss: What They Add to the Conversation between Women Theologians

What does a cosmology based on action, rather than substance, offer to the discussion of particularity and solidarity? Specifically, how will this enrich dialog between Mujerista, Womanist, Asian feminist and Feminist theologians? In general, an acknowledgment of self agency in relation supports our efforts at solidarity through simultaneously affirming both our individual and relational value. In our relationality, we can contribute to one another's history and current process of becoming. We can support, uplift, or hamper others' freedom and creativity in these relations. In our individuality, we can welcome the contributions of truly distinct voices, seeking for novel responses to old patterns of thought.

Defining women in terms of agency accomplishes two objectives. First, it replaces static boundaries with a definition of the individual based on creative self agency. Physical boundaries, then, are defined in the context of personal agency, nurtured or violated. Practically, this is important, because it provides one logical link to better understanding why physical violation of women's bodies is experienced in terms of a loss of agency. An agential definition of selfhood in relation is also important to feminism so that women are not defined only in terms of their relational value. Instead, women are defined in terms of their abilities as creators and shapers of themselves and their world. In recounting history, in crediting many major accomplishments of humanity, it is precisely this agency of women that has been ignored or denied. Women can be encouraged in their acts of defiant defining, of themselves and the overall social structures in which they participate. They may draw from both communal and personal resources in order to achieve more just relationships.

Second, even while we are internally related, others cannot make decisions for us, and our own creative ability can never be entirely negated. In a process metaphysic, the responsibility for decision and action is ultimately based on the agency or action of individuals. These decisions accumulate for good and for ill and are inherited in the present as a whole social fabric, but this fabric is built thread by thread through subjective decisions in the past. What this means in terms of human-human, human-divine, and human-nature relations is that an individual, while deeply in relation, is responsible for her own integrating or negating response to the influences around and within her. . These decisions, of course, are never done in complete freedom from the past, and others' actions may severely limit our possibilities. But, although an individual's creative self agency may be severely damaged by the influence of others, it cannot be totally squelched. Even within systems of great oppression, voices of creativity and resistance can rise to the surface since others' actions to influence us are not the last word.

In addition to this affirmation of women's agency, an emphasis on particularity or distinction acknowledges that true difference between individuals does not sabotage togetherness. Women of different races, social classes, and backgrounds may maintain their own integrity rather than being dissolved in the overall conversation with other women. The assertion of individual voice and agency, in this context, is not seen as oppositional or destructive to community but, rather, as constructive and integral to its very foundation. Capturing this perfectly, Howell writes, "Contrasts make possible the unity of diversity within experiences."[34] In another application of encouraging women's individuality, we can encourage women who are committed to community and justice to divert important energy to fostering their own health and becoming, whatever this entails. If commitments to teaching, to scholarship, to community activism, for the sake of further justice in our communities, lead women to neglect their own need for creative space, this may ultimately hinder, rather than benefit, others. As noted earlier, Howell mentions that some Womanists have emphasized that issues of "self" are not integral to their work for the good of the whole community, while white feminists have perseverated on it. From a process perspective, however, the good of the community and the good of the individual are not separate but intricately connected goals. Viewing the aim of justice from the lens of the community tames our society's individualistic emphases. Viewing the aim of justice from the lens of particular individuals emphasizes the value of singular or small aspects within the larger context. With both lenses working together, we can gain depth perception.

[34] Howell, *A Feminist Cosmology*, 94.

Furthermore, in arenas where true difference is welcome, conflict can become a source of growth. As stated earlier, women's solidarity is advanced and deepened by acknowledging great beauty and strength in contrast. Howell underscores the value of contrast because it encourages people to bravely engage difficult and fruitful conversations with others whose experience challenges one's own. In the meeting place of these perspectives, no matter how conflictual, one encounters beauty. Whitehead's very definition of beauty entails that an increase in beauty equals an increase in contrast within one individual occasion of experience. By his definition, beauty is the ability to hold within each moment, within each organism, many possible propositions for being. A certain flower may display a pink color, but hold within it the possibilities of several other colors, shades or shapes. The higher the intensity and contrast of possibility within a moment or organism, the more beautiful it is, according to Whitehead. Beauty enjoys contrast; and when there is more complexity, more intensity, there is more life. To live a life of beauty is to seek the highest influx of contrast, "freshness" and "zest" in each moment.[35] In fact, to seek life at all is to seek Beauty because the opposite of beauty is the inhibition of possibilities, which leads to atrophy, decline, or death.[36] Thus, as we are intentional in our efforts to magnify the intensity of our interrelatedness through dialog, we engage beauty and life even in difficult or painful conversations.

Last, in examining the particularity of each occasion of experience, one may appreciate unique individuals or moments within the ephemeral fleetingness of life. While we are all interrelated, each person and being is distinct. Moments and individuals are unrepeatable in agency and perspective. Even as past events and people may have a lasting effect, the exact form of creativity they offered ceases to exist. As Hazel Dickens noted in the opening of this paper, there are particular, unrepeatable loves. Even within the life of one person, one knows the swift pace of time and the perpetual shifts in perspective as one moves through life. While there is much inherited from the past, creating a continuity of personality, there are also drastic and subtle changes within one's life path and self. Change entails both addition and negation or loss. In Whitehead, even what is "negatively prehended" or not incorporated in the present leaves a scar, and thus we carry loss with us continuously. Whether we discuss moments or individuals, there is real loss even within a world of webbed relations.

---

[35] Whitehead, *Adventures in Ideas* (1933; reprinted, New York: Free Press, 1961) 258.

[36] Ibid., 259.

Recognizing the preciousness of each individual, we can renew our vigor in protecting one another and allow each other room to grow and speak honestly, in ways that support and challenge us. In addition to this, we gain a sense of honesty and freedom. We are honest in stating that we are interrelated, in this way we hold the past within us and all is not lost; and yet we also know true loss and death, and we are free to mourn and face the deaths, both large and small, we experience throughout our living.

3

Story, Forgiveness, and Promise:
Narrative Contributions to a Feminist Cosmology

Lisa Stenmark

In *A Feminist Cosmology,* Nancy Howell develops a constructive feminist theory of relations,[1] heavily influenced by the process philosophy of Alfred North Whitehead. Blending process and feminist thought, she argues for a cosmology that focuses on women's experiences, incorporates a wide variety of perspectives, and is committed to particularity and praxis.

Howell is right to assert that process and feminist thought each benefit from the insights of the other, and her book marks a tremendous contribution to both. The usefulness of her synthesis is evidenced in her discussions of nature, relationship (between persons), and the divine, which clearly delineate the issues, ask hard questions, and chart a tentative route through the territory.

This paper will expand on the project that Howell has begun by adding a narrative thread to the tapestry. I will begin by introducing two approaches to narrative thought, one of which addresses narrative as arising from temporal experience and the second of which focuses on narrative as a product of human action.[2] This will demonstrate a basic compatibility between narrative thought and Howell's approach. In the second

---

[1] Nancy R. Howell, *A Feminist Cosmology: Ecology, Solidarity, and Metaphysics* (Amherst, N.Y.: Humanity, 2000).

[2] These approaches are not exhaustive of narrative theory. For example, although I include Stanley Hauerwas in the first approach, his understanding of narrative also represents a third approach, post-liberalism, which understands narrative as related to culture and tradition.

part, I argue that narrative is necessary to an experiential approach such as Howell's and then explore what narrative can contribute to a theory of community and praxis. In so doing, I will also outline some of the criteria for evaluating narratives.

Narrative as Temporal Experience

This first approach includes such thinkers as Paul Ricoeur, Stephen Crites, and Stanley Hauerwas. This approach focuses on narrative as it arises from an experience of time as a continually perishing present in relation to the past and future: the past is past, but continues through memory; the future is not yet, but is already through anticipation; the present is fleeting to the point of non-existence, but is held by our attention to it. These three aspects of consciousness—memory, attention, and anticipation—organize temporal experience into a single, coherent whole, which is expressed through narrative or story. It is through narrative—and narrative alone—that human beings express and configure temporal experience. It is through narrative that we are able to bring meaning to—or reveal the meaning in—our lives.

This approach has clear resonance with process. But, unlike process, which clearly purports to describe the structure of existence, a narrative trajectory begs the question of whether narratives are merely a human construct or whether they are the way the world really is. Hauerwas, for example, rejects any ontological claims for narratives.[3] Ricoeur, on the other hand, describes unmediated experience as an "untold story" from which "told stories" emerge. Human beings are entangled in stories; told stories are merely "a secondary process, that of 'the story's becoming known.'"[4]

One reason for the confusion about the status of narrative is disagreement on the meaning of the term. Clarification of its possible meanings is suggested by Calvin Schrag's distinction between "stronger" and "weaker" narratives: stronger narratives are "expressive of an ontological claim," while weaker narratives describe a form of discourse and a textual understanding.[5] His distinction is flawed, however, because Schrag describes

---

[3] "The structure of a particular narrative is not in any exact way the structure of reality, but narrative form is a necessary way of seeing, transforming, and, to some extent, re-expressing reality true to the form of human activity." Stanley Hauerwas, "Story and Theology," in *Truthfulness and Tragedy: Further Investigations in Christian Ethics* (Notre Dame: University of Notre Dame Press, 1977) 76.

[4] Paul Ricoeur, *Time and Narrative* (Chicago: University of Chicago Press, 1983) 1, 75.

[5] Calvin O. Schrag, *The Self after Postmodernity* (New Haven: Yale University Press, 1997) 42.

stronger narratives both as ontological and as descriptive of the human self.[6] Since these are distinct understandings,[7] I distinguish two stronger forms of narrative.[8]

I understand "narrative" as encompassing three distinct but interconnected "kinds" of stories. There are the "untold stories," which describe the temporal structure of existence; the narrative self, which is human consciousness and identity as it exists through time; and the stories we tell. The connection is consciousness, which mediates between the world and our stories. Consciousness is shaped by the temporal structure of the world and, in turn, expresses and shapes human experience of the world.

Our initial experience is the world as it is, lived time, the untold story. Although some experiences might be better expressed in non-narrative form, the particular manifestation of lived experience—existence in and through time—is narrative. Critical reflection and abstract insight remain necessary components of our conscious mediation of the world, but it is through narrative that persons appropriate insight and reflection into their experience of the world.

Narratives are able both to express time and incorporate insight through emplotment. Plots combine multiple events and factors (agents, goals, interactions, results, etc.) along with insight, reflection, and evaluation. They are the "privileged means by which we re-configure our confused, unformed, and at the limit mute temporal experience."[9] Through emplotment human beings create a configuration by picking and choosing, bracketing, organizing, and shaping unconnected and open-ended events, thus transforming experience into a coherent whole.

By choosing to tell a story in a particular way and from a certain perspective, perhaps even offering a parenthetical explanation, we give a particular meaning to an event. Narratives allow us to reflect on and describe our experiences, so that reflection and perception become a part of experience. As we continue to reflect on our experience, our stories pass

---

[6] Ibid.

[7] It would be possible, for example, to view human personal identity or human perception as narrative, without saying that narrative describes the world as it actually is, as, in fact, Stanley Hauerwas does.

[8] Crites speaks similarly of "sacred stories," which not only lie deep in the consciousness, they "form consciousness rather than being among the objects of which it is directly aware." Stephen Crites, "The Narrative Quality of Experience," in *Why Narrative? Readings in Narrative Theology*, ed. Stanley Hauerwas and L. Gregory Jones (1989; reprinted, Eugene, Ore.: Wipf & Stock, 1997) 69.

[9] Ricoeur, *Time and Narrative*, 1, xi.

through consciousness a number of times. This blend of "how the world really is" or "what really happened" and critical insight forms an hermeneutical circle, as the meaning evolves and deepens when the story is told and retold. This circle is not vicious, returning again and again to the same place, but is instead "an endless spiral" that continually returns to the same point, but (ideally) at different altitudes.[10]

Because of the relationship between experience and reflection, narratives do not simply duplicate the world. They allow us to see the world more clearly, augmenting it with meaning. Just as metaphors create meaning by revealing previously unperceived relationships and drawing discordant meanings into a concordant whole, narratives create meaning by drawing together previously unconnected events and experiences. A narrative frames and configures reality, bringing discordant meanings and events into a concordant and meaning-full whole and revealing a relationship that ordinary vision does not reveal.[11]

In addition to revealing a depth of meaning in the world, narratives also create possibilities. Because narratives allow us to "rearrange" the world in endless ways, they become "an immense laboratory for thought experiment," allowing us to try on new roles, new actions, and new possibilities without the risk of actualizing them.[12] Through this virtually unlimited power to configure novel ontologies, narratives disorient and thereby reorient the world by disorienting and then reorienting consciousness. They allow us to see things we never saw before, change our perception, and thereby change our consciousness. Narratives therefore change the world, in part, because they change consciousness. As process metaphysics suggests, this is more than "just" a mental change. Because consciousness is a part of the world, a change in consciousness changes the world.

For those unwilling to accept process metaphysics, narratives change the world on a more practical level. By changing our perception of the world and enabling us to imagine alternatives, narratives cause us to act differently. You cannot do what you cannot dream, and narratives are how we dream. Narratives are vital to developing a vision of a world, which we can then choose to create.

Because of striking similarities between process and narrative, and thus between narrative and Howell's feminist cosmology, her conceptual

---

[10]   Ibid., 72.

[11]   Ricoeur, *Interpretation and Theory: Discourse and the Surplus of Meaning* (Fort Worth: Texas Christian University Press, 1976) 67.

[12]   Ricoeur, *Oneself as Another*, trans. Kathleen Blamey (Chicago: University of Chicago Press, 1992) 89.

framework is exceptionally well suited to adding depth of reflection to our stories.[13] But stories, no matter how deeply influenced by conceptual insights, do not create meaning unless those insights become incorporated into experience through our stories. Stories help actualize the creative potential suggested by Howell's cosmology. They allow one to test insights, to configure, reconfigure, and even re-reconfigure the world. The full realization of creativity requires both insight and story, because there is no reconfiguration without analysis and no actualization without stories.

Narrative as Speech and Action

A second approach to narrative focuses on narrative, not as it arises from the experience of time, but as it emerges through human action,[14] an approach typified by Hannah Arendt. In *The Human Condition*, she describes the highest form of human activity as action.[15] This is not merely doing something, but doing something in such a way that we reveal who we are. For action to have this revelatory quality it must be tied to speech. Speechless activity is not action, because without speech, there is no actor. "The action . . . becomes relevant only through the spoken word in which [the speaker] identifies himself as the actor, announcing what he does, has done, and intends to do."[16]

In order for us to be persons, action and speech are both necessary and must be in consonance with one another. Through action, we insert ourselves into the world and initiate beginnings; through speech we reveal that action as ours and ourselves as agents. In other words, for my actions to reveal who I am I must be able to tell my own story.

[13] For example, memory, attention, and anticipation coincide with the dynamics of an actual occasion. Through emplotment, narratives present novel possibilities; and through concrescence, actual occasions exercise their creativity and form novel configurations. This suggests that narratives do not simply impose meaning on events, but that meaning is an intrinsic property; that emplotment does not merely introduce meaning into events, but may reveal a meaning that is already present in the events.

[14] While less obvious in its connection to process and thus to Howell's cosmology, this second approach is connected because of its connection to the first approach. Temporality and activity are not unrelated. Action occurs in and through time (however briefly). The two trajectories examine the same human condition from different directions. Thus, the temporal approach addresses the experience of temporality, but stresses the need for action, and the action approach addresses the experience of time. This second trajectory is important because, as I will demonstrate in the second part, it has significant contributions to make to Howell's project, especially on the issues of community and praxis.

[15] Hannah Arendt, *The Human Condition* (Chicago: University of Chicago Press, 1958).

[16] Ibid., 178–79.

Stories are more than a framework for action, stories are the glue that bind us to other human beings through action. Human action creates a story because all action takes place within the "web of human relationships," which exists because we are born into a world of already acting human beings. On the surface, speech and action deal with tangible objects that we can see, touch, etc. Every time I say and do something, it creates a relationship with innumerable others. This is a "second, subjective in-between," which is an intangible, but real connection.[17] When I act and speak, I do so in a context of a world of other human beings who are affected—directly and indirectly—by my speech and action. Speech and action both create the unique story that is mine and affect the stories of all of those who with whom I come in contact. Their (re)actions contribute to my story, even as my (re)actions contribute to theirs. It is because of the web that actions produce stories, "as naturally as fabrication produces tangible things."[18]

This suggests that action and speech have a two-fold character: it is initiated by one but completed by many. Because action involves the "innumerable, conflicting wills and intentions" of countess others whom we cannot control, even by force, an individual may initiate an action, but no individual—not even a leader—can complete an action, because the moment an action is initiated it enters into the web and is taken in by others, who will do with it what they will.[19]

As a result, action is unpredictable. Persons unknown and unforeseen can respond to what I have done in a complex, endless, and therefore exponentially unpredictable way. I can never ensure that my actions will achieve fulfillment in anything resembling my original intent. Human beings are helpless to bring their actions to fulfillment without the help of others, and. for this reason, action almost never achieves its purpose.

Unpredictability combines with irreversibility, the basic reality of linear temporality in which it is impossible to undo what we have done. Once an action is initiated, it cannot be undone, no matter how much it is later regretted. Persons remain responsible for their actions, even if the outcome was not their intent. This has devastating consequences for actions within the web, because they can go on forever: "The process of a

---

[17]   Ibid., 183.

[18]   Ibid., 184.

[19]   Ibid., 188.

single deed can quite literally endure throughout time until mankind itself has come to an end."[20]

There is an "antidote" to the unpredictability and irreversibility of action and speech, namely, two special kinds of action and speech: forgiveness and promise. Both are necessary for the fulfillment of action and, therefore, for the emergence of human beings and the formation of communities.

Without being forgiven, released from the consequences of what we have done, our capacity to act would, as it were, be confined to one single deed from which we could never recover; we would remain the victims of its consequences forever, not unlike the sorcerer's apprentice who lacked the magic formula to break the spell. Without being bound to the fulfillment of promises, we would never be able to keep our identities; we would be condemned to wander helplessly and without direction in the darkness of each man's lonely heart, caught in its contradictions and equivocalities.[21]

Forgiveness has the power to free us from the irreversibility of action. "The possible redemption from the predicament of irreversibility—of being unable to undo what one has done though one did not, and could not, have known what he was doing—is the faculty of forgiving."[22] Forgiveness is unique and utterly creative. It "is the only reaction which does not merely re-act but acts anew and unexpectedly, unconditioned by the act which provoked it and therefore freeing from its consequences both the one who forgives and the one who is forgiven."[23]

The second solution is promise. It is the only alternative to domination, which is the attempt to control human unpredictability, forcing others to respond in a set way to initiated action. Through violence, it is possible to eliminate unpredictability, but this also destroys spontaneity. The alternative is promise: "binding oneself through promises, serves to set up in the ocean of uncertainty, which the future is by definition, islands of security without which not even continuity, let alone durability of any kind, would be impossible in the relationships between men."[24] Unlike domination, promise leaves human spontaneity in place. It releases the

---

[20] Ibid., 233

[21] Ibid., 237.

[22] Ibid.

[23] Ibid., 241.

[24] Ibid., 237.

power generated when we come together by enabling us to act in concert in creative ways.

A Feminist Story of Creation

The preceding discussion clearly illustrates the compatibility between narrative approaches and Howell's feminist cosmology. Because of this compatibility, Howell's approach in *A Feminist Cosmology* is well suited to reflecting on stories. But the relationship is not one way. For, while Howell's approach is promising, it fails to make the connection between theory and action (or thought and practice) that is called for by a commitment to praxis. What is needed—as I argue in the first section—are stories, because it is only through the structure of narrative that it is possible to integrate insight and reflection into "who" we are and what we do. In this section, I will highlight several issues that Howell raises, specifically, the need to frame particular experiences and the issues of community and praxis, in order to explore the contributions of narrative insight.[25] In so doing, I will consider some of the characteristics of a feminist story of creation.

Telling Our Story:
Experience, Particularity and Inclusively

Experience, as Howell points out, is important for both process and feminist perspectives. This commitment to experience makes narratives indispensable, because narrative is the privileged means by which human beings explore their experience of the world. Experience is irreducibly temporal, and only narrative "can contain the full temporality of experience in a unity of form."[26] Narratives are the only non-abstract way to speak of lived experience. Narrative is further privileged in a cosmology that is relational and organic, because narrative makes a "connection among elements (ac-

---

[25] These are issues that Howell raises both in her outline of the principles of a feminist cosmology (chapter 2) and in her fourth chapter, "Relating to Each Other." This should not suggest these are the only compatibilities between Howell's approach and narrative. For example, her discussion of nature in the third chapter is an important corrective for narrative thought, which almost universally ignores nature. Process reminds narrative that he story of God's work in the world is not a story with only two characters, God and human beings. It is a story involving all of creation. Narrative is not without contributions, on this issue, however. Arendt offers a caution about human intervention in nature. Because in our relationship with nature there is no forgiveness and no promise, there is no way to stop or control the effects of human action within that web. While Arendt's observation is not without flaws, it does raise interesting questions.

[26] Crites, "Narrative Quality," 78.

tions, events, situations), which is neither one of logical connection, nor one of sequence."[27]

Narratives are also indispensable to the tasks of including a wide range of experiences and focusing on the subjective experience of particular women (i.e. the poor), two feminist and process commitments that Howell affirms. Both narrative trajectories outlined above include the admonition that persons must tell their own stories. This is clear in the assertions that self-hood is connected to our ability to tell a coherent story and that the emergence of stories is dependent upon speech and action being connected. When a "story" is told by others, speech and action are less likely to be connected. That is, it becomes less likely that the words reveal the person who is acting.

Consider a woman—she happens to be Asian, perhaps Vietnamese (she sometimes wears a conical hat)—who collects cans and bottles on and around a major metropolitan university. Watching her, it is easy to construct a story of her life: the ravages of war, the scourge of poverty, the alienation of forced exile in a foreign country. These things—collectively or alone—have reduced her to rooting around in garbage. I am committed to the proposition that theological reflection must arise from and resonate with this woman's experiences. A noble expectation, but narrative theory raises cautions about telling someone else's story.

As it turns out, this woman—and her husband—lost their jobs several years ago. They were determined to do whatever took to thrive in a place where they believed that dreams could come true, for themselves and their children. They began collecting cans and bottles to tide them over, but discovered that with hard work this could earn them a decent living. Now, almost ten years later, they have put two kids through college (one now in graduate school) with a third ready to finish.

The difference between the story I told about this woman and the story she tells about herself is that only her story reveals who she is. In my story, the woman was a victim. In her story, she was decisive and adaptive, creating her own future by initiating stunning new beginnings. In short, only her story reveals her as an agent, or a "who."

Narrative does more than admonish us to allow persons to tell their own story; narratives themselves make it possible for us to see the world through the eyes of others. Stories make it possible to know another's experience of the world, even when that experience is radically different from

---

[27] Hauerwas, "From System to Story: An Alternative Pattern for Rationality in Ethics," in *Truthfulness and Tragedy*, 28 [15–39]; reprinted in *Why Narrative?*

ours. Stories permit a glimpse of the world through another's eyes and allow us to live for a moment in someone else's skin.

Narratives also help resolve a tension inherent in the seemingly conflicting demands for particularity and inclusivity: the expectation that we must look through the eyes of particular (marginalized) persons while simultaneously including a wide range of perspectives and rejecting universal claims. This tension is present in Howell's work. She rightly points out that if we do not demarcate our experience, we risk universalizing our perspectives.[28] Yet she also asserts—again rightly—that white feminists separate themselves from women of color when they demarcate their experience.[29]

The contradictory demand to particularize and include can be resolved by narratives, which allow us to approach a story from a particular perspective while including a wide range of views. A single narrative can look at the world from multiple perspectives, combining them into a whole, demonstrating how they intertwine, without privileging any one or even resolving conflicts within those perspectives. But not all perspectives are equal. Narratives make it possible to frame a story in a certain way, even while including multiple perspectives. This is similar to what Howell calls "importance," or the organization of data in a particular configuration. In narrative language, this can be thought of as the overall "plot" that guides us in putting the story together.

A narrative approach also raises cautions about the admonition to let persons tell their own story. First, there are circumstances, such as extreme suffering, that leave persons mute. In those cases, it is necessary to give voice to the voiceless. As demonstrated above, this is a tricky proposition and must be done with extreme care. Even then, the story will be filled with distortions, so that stories told for others involve intensely hermeneutical problems. The second caution is a human tendency to self-deception. Even when persons tell their own story, they lie to themselves. In other words, even when we tell our own story, there are hermeneutical issues.

Because stories are hermeneutical, a simple admonition to tell stories is not enough. We clearly need a way to evaluate our stories and the stories of others. But, in a postmodern world, it is difficult to defend abstract, independent criteria. Howell, for example, suggests "the unity of the universe" as a criteria for organization.[30] I do not disagree with the principle,

---

[28] Howell, *Feminist Cosmology*, 17–18.

[29] Ibid., 86.

[30] Ibid., 18.

but it is difficult to support, unless one is committed to process meta-physics (or feminist principles). This assertion, like all assertions, is deeply rooted in a particular narrative.

Ultimately, the only way to evaluate stories is with other stories. Or, more specifically, with a larger story, a kind of cultural narrative: the history, traditions, evolution, etc. of a particular community. Commitment to a cultural narrative or tradition is antithetical to a modernist view, where truth lies in abstract principles. In the vestiges of Modernism, the suggestion that I speak "as a Christian" excludes rather then includes. But in a narrative view abstracts do not include, and particularity does not exclude.

My speaking as a Christian while others speak as Hindus, Buddhists, or atheists actually opens up space for conversation. Abstract, "reasonable" assertions exclude because, if we disagree, one of us is "unreasonable" (and it ain't *me*). But when disagreements are based in a particular tradition, they are based in our different stories. This makes it possible to talk about and understand real differences and why they exist. It becomes possible to make informed, thoughtful decisions about which particular perspectives or stories are the most appropriate for a particular task.[31]

The criteria for evaluating narratives must also lie in our understanding of narratives themselves. That is, do those narratives do what they are supposed to do? The immediately preceding discussion implies that one of the criteria is whether it is a story told by someone about themselves—whether it reveals who they are—rather than a story told to or about them. This suggests a commitment to empowering the voiceless to speak or, in extreme circumstances, telling their story for them.[32]

Community and Forgiveness

Both Howell's feminist cosmology and a narrative view include the assertion that there are no individuals distinct from relationships. As Howell points out, process metaphysics reinforce this perspective in the doctrine

---

[31] The lack of a narrative is a drawback to a process approach. Absent a narrative, it will remain an academic discourse and will not be able to accomplish its vision. The task for process theologians will be to translate process notions into stories and to incorporate them into the broader cultural narrative. Feminism too lacks a narrative. This is the root of the ongoing struggle to define just who it is that feminists speak for. Absent an identifiable community, the feminist narrative ultimately loses coherence.

[32] The necessity of speaking for the voiceless is not restricted to human beings. Animals, plants and creation itself are often voiceless, and Howell makes it clear that we need to think about how to tell their story, and include it in our story.

of internal relations, describing everything as internally related to everything else. Neither process thought nor Howell's approach makes the connection between this relational view—individuals in society—and understanding individuals in community. Her feminist cosmology has a theory of relationship but does not appear to have a theory of community. Narrative contributes to the development of such a theory.

In a narrative view, individuals and communities are inextricably connected. "The self that is called into being through discourse and action is at the same time called into being within a community."[33] The same action and speech that forms us as persons binds us to one another. In order to be, we must act and speak; but whenever we act and speak, we are drawn into relationships. We cannot be without being in community; we cannot be in community without being persons. All that persons think, say, and do happens in and because a community sets the boundaries of possible thought and action. A community creates a context in which what a person thinks and how that person acts makes sense. Absent a community, there is no person to act and no context in which to act.

Space considerations preclude a full exploration of a narrative theory of community, but I will demonstrate the value of this approach by examining an issue that Howell raises, which is a tension between White feminists and women of color. Howell argues that the unexamined racism of White feminists can—and often does—preclude a sense of community with women of color. I believe that a narrative approach sheds light on how to address this issue. Narrative theory, and actual narratives, suggest that through confrontation, transformation, and forgiveness, it is possible to reconcile and (re)form a community.

In *The Color Purple*, Alice Walker describes a process of transformation that transcends a history of pain, betrayal, and even abuse. Many of the main characters have hurt one another—Celie tells Harpo to beat Sofia, Mister beats and abuses Celie, Eleanor Jane benefits from a system built on Sofia's suffering. But by the end of the book, these people have reconciled, and some have even developed nurturing relationships.

In each case, reconciliation begins with a confrontation, when those who have been injured find their voice and, through their stories, the offenders are able to see what they have done and the pain they have caused. Stories are a vital part of confrontation. By linking persons, action, motivations, and results, they connect us to our actions and their effects. They

---

[33] Schrag, *Self,* 77.

also allow us to see our actions through the eyes of another, thus confronting us with our deeds.

Reconciliation, therefore, begins with the admonition that people must tell their own stories and that they must be empowered to do so. But there must also be a commitment to hear. Transformation—a changing of ways—only becomes possible after offenders see what they have done and accept responsibility for their actions. The experience of seeing by looking at the world through the eyes of another—if one is truly open to that experience—can be both disorienting and transformative.

Seeing ourselves as others see us can be painful. But women—particularly those who are privileged—must be willing to be confronted by stories in which we are neither hero nor victim. Walker makes it clear that at least some of us are Miss Eleanor Jane,[34] who wants acceptance from her former maid Sofia, but is unwilling or unable to see how she has benefited from Sofia's suffering. It is only when she is confronted by Sofia's story—that Sofia does not love Eleanor Jane's son—that she begins to ask hard questions. Once she asks and starts to accept the answers, she begins to change. It is her change that is the catalyst for Sofia and Eleanor Jane to embark on the slow process of reconciliation.

It is not the responsibility of women of color to forgive the past; the responsibility for reconciliation lies with privileged women, who must be willing to change. But this does not mean that forgiveness is unimportant. Walker does not address the question of what it takes to ask the kind of questions that lead to change. For those of us who are privileged, the answers to hard questions—What is housing like for those who pick my artichokes? Who lives where my garbage gets dumped?—not only confront us with the need to change, they raise the possibility of a guilt from which we can never recover because we cannot undo do the past, no matter how much we might regret it.

The only way to bring myself to face my guilt is through the possibility of forgiveness, because forgiveness alone marks the human capacity to do and begin something new, unencumbered by the past. The act of forgiving frees us from the past. In the language of narrative, it reconfigures the past in a way that leaves us open to a new future. It is essential to transformation, because it frees the guilty from the impossible burden of

---

[34] Miss Eleanor Jane is the daughter of the mayor, whom Sofia was forced to work for. As a result of her forced servitude, Sofia literally never saw her own children. Eleanor Jane, now an adult continues to seek Sofia's approval and in the final confrontation, has brought her son, Reynolds Stanley, to Sofia's house and is trying to get her to say that she loves him.

making up for what they have done. Forgiveness is essential to meaningful communities, because it makes true reconciliation possible.

Confrontation, transformation, and forgiveness are not a step-by-step formula for reconciliation. Not all confrontation leads to change. There are those who, when confronted, refuse to own up to their actions and do not become part of the community (i.e., Celie's father). And it is possible to forgive someone who has not changed, just as it is possible to change without being forgiven. But reconciliation needs both transformation and forgiveness. Refusing to take responsibility alienates speech and action, forcing us to live an untrue story. Those who refuse to confront their past cannot be transformed. They are less than fully human and less than fully communal. As Sofia points out, it does not matter to her whether or not Eleanor Jane changes, because it is not Sofia's salvation that Eleanor Jane is working for.

Forgiveness is also vital, because it is the hope of forgiveness that makes it possible to confront our past and thus to change. Forgiveness is also necessary, because a community must be built on a shared past, a past that will always include suffering and misdeeds, whether by ignorance or malice. Without forgiveness, we would never be able to do anything new and would be forever doomed to relive a broken history

Because the criteria for evaluating a story lie in the characteristics of narrative, a story must be one that creates community. This reaffirms the previous criteria for evaluating stories. First, a story is told by someone and not about them (everything changed when Eleanor Jane stopped telling Sofia's story); and, second, we must be committed to giving voice to the voiceless (the confrontations in *The Color Purple* begin when someone who has suffered in silence begins to speak). In addition, this story must enable us to see the truth about what we are doing. That is, our words must match our deeds. A story that allows us to lie abut ourselves—or that encourages it—is not a story that creates community. Finally, it needs not only to be a story in which forgiveness makes sense, but also a story that encourages forgiveness and, therefore, reconciliation and community.[35]

---

[35] A fuller narrative discussion of confrontation, forgiveness, and reconciliation would include the meaning and implications of the story of forgiveness that forms the basis of the Christian narrative and the Christian community. As a Lutheran, I would like to suggest that the tradition's emphasis on Law and Gospel might also shed light on the topic.

Promise and Praxis

Howell argues for an approach that allows us to reflect on experience and inspires praxis.[36] I agree that any theological approach worth its salt does both. As stated above, her approach provides a valuable tool to reflect on experience, but its lack of a theory of community means that her approach cannot adequately theorize praxis, describing how a community's commitments can be translated into action. Because narrative has a more developed theory of community, it has a more developed theory of action, and, therefore, of praxis.

If I want to change the world, I cannot act alone. Or, more accurately, I can act alone; but for that action to achieve a desired result, there must be others willing to act with me, willing to commit to a particular course of action. Effective action requires a community, and, just as importantly, in a narrative view a community requires action. Just as I must be an actor in order to be a person, so too must communities center around and arise from action, both from the action of an individual, which draws others in through the web, and through shared action.

Shared action requires people who can act in concert. This is only possible through promise, by which persons commit themselves to a particular course of action. Promise overcomes "the impossibility of foretelling the consequences of an act within a community of equals."[37] It creates the opportunity for shared action by creating a "space of appearance," which is "the power generated when people gather together and 'act in concert.'"[38] It is in the space of appearance—and only there—that human beings are able to achieve power.

The actualization of power is dependent on mutuality. This relationship reflects a connection between our stories and our communities. Communities require that words and deeds connect. When we lie about who we are and what we are up to, we cannot form communities and cannot exercise the power of mutuality. When we are isolated from one another—by choice or by force—speech and action are no longer possible, and we can no longer freely make promises. Action loses its power when togetherness is lost, in those times when we are acting merely for or against, but not with others, "speech becomes indeed 'mere talk,' simply one more means to an end."[39]

[36] Howell, *A Feminist Cosmology,* 19.
[37] Arendt, *The Human Condition,* 244.
[38] Ibid., 244.
[39] Ibid., 181.

The power of communities does not replace the responsibility of persons. The power of action is closely tied to the disclosure of the agent, to the ability of persons to say who they are and what they will do. In order for communities to continue to act and in order for them to from around shared action and shared promise, there must be persons capable of acting and of making promises. We must be willing to name ourselves in the community and bind ourselves to action. Mutual promise is a way to get the benefits of concerted action while retaining individual responsibility and spontaneity

The purpose of narrative is to form communities, and communities need to form around shared action. Therefore, another attribute of a good story is that it encourages promise making and promise keeping. Further, the stories that we need to tell are those that inspire us to act, and they must configure a possible future.[40] Finally, because of the difficulty in achieving action, the story must sustain us through hardship and suffering, inspiring courage and hope.

At the November 2000 Annual Meeting of the American Academy of Religion, bel hooks was speaking of the American Civil Rights Movement, and the deaths of James Chaney, Andrew Goodman, and Michael Schwerner (killed in Philadelphia, Mississippi, on June 21, 1964). She asked: "Do you remember when people loved justice enough that they were willing to die for it?"

It is not an easy question. Even harder is the suspicion that the discourses of American Culture in general—and American religious discourse in particular—lack the vision and the power to make us love justice or be willing to suffer, much less die, in order to achieve it. Unfortunately, Howell's feminist cosmology is not such a story. But, then again, neither is a narrative approach. A commitment to metaphysical concepts or academic theories—even those as persuasive as process or narrative—cannot be the basis of community. Neither a cosmology nor a theory can make us hopeful or inspire action. What we need are stories.

The year 2002 saw the anniversaries of several landmarks of the civil rights struggle, and there were several programs commemorating these events. In one, a woman—now in her late 50's—recalled being a young girl integrating a grade school. What she remembered was that none of the other kids would play with her. What she remembered was that no one to would jump rope with her. As an adult, more than fifty years later, she began to cry, remembering the terrible loneliness of the struggle for justice.

---

[40] Forgiveness deals with the past, promise deals with the future.

Then she composed herself, recalling that at the time, it did not seem so bad. We all knew, she said, that God was on our side. And He would never abandon us, and would never give us more than we could bear. She was able to endure, because she lived in a community inspired to love justice enough to suffer, a community sustained by the story of a loving God who would not abandon them.

Academic pronouncement of praxis must do several things, all of which center around a lonely little girl with faith in a God who would not abandon her. It must give us a vision of a world of justice, it must give us hope that our vision is not empty, and it must sustain us in the struggle to achieve that vision. It must contribute to stories that give rise to communities of people who love justice enough that they are willing to die for it. In the words of Stanley Hauerwas, it must help us to go on. In the words of Dolores Williams, it must make a way out of no way.

There are, of course, many such stories. For many, the life, death, and resurrection of Jesus and the ongoing life of the Church have been that story. But transcendent and transformative stories have lost their sway in Western culture—and the academic study of religion certainly bears part of the blame for this.[41] Our most powerful story of transcendent hope in the midst of darkness—the breathtaking news that God became flesh and dwelt among us—has been reduced to a tale about whether retail stores made more money this year than last year. Our stories have become shallow, lacking the depth and power to bind us to anyone or anything, to inspire us to forgive, or to bind us with a promise.

Neither a feminist cosmology nor narrative theory can replace the stories that we have lost. They cannot create a people who love justice, and they cannot sustain a lonely little girl. But they can provide the basis of a critique of the demonic structures that hold sway in our culture, and they can allow us to see the world in ways that will enrich and deepen our stories. The challenge will be to incorporate those insights into our stories, so that those insights can inform our lives and our actions, binding us together as a people who love justice and who have the strength to make a way out of no way.

---

[41] A discussion of how it is that stories have lost their meaning and power—and the devastating effects of that loss—are an important component of the postliberal approach to narrative. The most well known would be the writings of Stanley Hauerwas.

4

Dualism without Domination:
A Reinterpretation of Dualism for Ecofeminist Theory

Kathlyn A. Breazeale

> Feminists have asked, Is dualism an adequate framework for un-
> derstanding reality? Overwhelmingly, feminists reject dualism that
> dichotomizes spirit and matter, culture and nature, humans and
> nature, mind and body in oppositional thinking that entails es-
> tablished assumptions about the superiority or superordinate value
> of spirit, culture, humans, and intelligence. Not only is dualism
> an oversimplification of the nature of reality, but dualism further
> exacerbates injustice toward women and nature.[1]

IN this quotation, Nancy R. Howell cogently articulates why feminists
have rejected dualism. Classical Western dualistic thinking conceives of
reality as composed of oppositional pairs in which one half of the pair is assumed to be inherently superior to the other half of the pair. Domination
of the superior half over the inferior half is therefore justified. Dualism
"further exacerbates injustice toward women and nature" because women
and nature have been assumed to be the inferior halves of the man/woman
and human/nature pairs. Howell also argues in the quotation above that
"dualism is an oversimplification of the nature of reality." While I do not
disagree with Howell's assessment of oversimplification, I offer an alternative approach to resolving the problem of domination caused by dualistic
thinking, an approach that does not reject dualism.

---

[1] Nancy R. Howell, *A Feminist Cosmology: Ecology, Solidarity, and Metaphysics* (Amherst,
N.Y.: Humanity, 2000) 48.

I hold that the "established assumptions" of superiority/inferiority or hierarchical thinking are the root of the domination of men over women and nature, not dualism *per se*.[2] I assert that dualism can provide a positive model for conceptualizing relationships in creation between entities that are distinct, but not separate, entities that are fluid and changing, entities that relate to each other without domination of one over the other. To support my position, I draw on three sources: selected feminist analyses of Hildegard of Bingen's theory of sex complementarity, Nahua (pre-Columbian Mesoamerican) concepts of duality, and Alfred North Whitehead's conception of the relationship between body and soul.[3]

In utilizing this diversity of religious and philosophical perspectives, I seek to contribute to the development of ecofeminist theory through an analysis of dualism that critically examines the pairs of man/women, culture/nature and spirit/matter. As Howell develops her feminist cosmology, she clearly describes how these three pairs have been central in justifying the domination of men over women and nature:

> Alienation of men from women and nature is based on the dualism of culture and nature and on the dualism of spirit and matter. These two dualistic pairs are repeatedly cited in feminist literature as the source of human and environmental injustice. Although it is not necessary that dualism create alienation, the culture/nature dichotomy and the spirit/matter dichotomy substantially support the valuation of men over women and nature and ultimately endorse the exploitation of women and nature.[4]

In the following sections, I discuss conceptions of duality that do *not* value men over women and nature, or spirit over matter. In contrast to Howell's definition of dualism as a view that "tends to overlook the option of the value and connection (or even the experience and existence) of both

---

[2] Howell defines "hierarchy" in part as "the pervasive rank ordering of exisitents in nature, such as the chain of being, God-angels-humans-animals-vegetation-inorganic matter. Hierarchy is rooted in cultural, sociohistorical, and often uncritical assumptions about value, complexity, and superiority measured with respect to human or abstract divine norms. Related to a logic of domination, hierarchy fuels racism, sexism, heterosexism, and other forms of oppression including exploitation of nature"; *A Feminist Cosmology*, 49.

[3] I am grateful to Karen Jo Torjesen for conversation and resources and to Vaughan McTernan for her essay "Performing God: God, the Organic and Postmodernism," *American Journal of Theology and Philosophy* 23 (2002) 236–51. These sources stimulated my thinking about dualism and ecofeminist theology.

[4] Howell, *A Feminist Cosmology*, 41.

poles of the dichotomized pair,"[5] I describe dualistic conceptions that posit a dynamic relationship between each half of the pair, a relationship that does not value one member of the pair over the other. I demonstrate how these conceptions of dynamic relationships in which each member of the pair is equally valued can undercut practices of domination and exploitation based on dualistic thinking. I conclude the conceptions of dualism without domination can be useful for ecofeminist theory to articulate how women *and* men are not separate from or above nature.

Hildegard of Bingen's Theory of Sex Complementarity

The medieval abbess Hildegard of Bingen (1098–1179) was a prominent theologian and advisor to many of the sacred and secular leaders of her time.[6] As a Christian nun, Hildegard inherited a religious tradition of asceticism that alienated women from their own bodies and from spirit and posited that women could only achieve spiritual equality with men in heaven.[7] Hildegard's theory of complementary sex dualism was in contrast to the theological anthropologies written before and after her work. For example, she rejected the traditional inequality of women and men, and she developed her theory a century before Thomas Aquinas quoted Aristotle in declaring that "the female is a misbegotten male."[8]

Prudence Allen, in her book *The Concept of Woman*, discusses how Hildegard constructed her original theory of sex complementarity based on the integral relationship of women *and* men with nature. Hildegard's theory is based in her belief that the four elements of nature—fire, air, water, and earth—are inseparable and humans live in relationship with these elements. Hildegard wrote:

> "Mankind [sic] lives out of the four elements. Namely, God has put the world together out of these four elements such that one cannot be separated from another; the world would no longer be,

---

[5] Ibid., 49.

[6] Hildegard's prominence and theology have been widely discussed by scholars in recent years. For two examples, see Matthew Fox, *Illuminations of Hildegard of Bingen* and Barbara Newman, *Sister of Wisdom: St. Hildegard's Theology of the Feminine*.

[7] Ruether, cited in Howell, *A Feminist Cosmology*, 45.

[8] Aquinas, quoted in Serenity Young, editor, *An Anthology of Sacred Texts by and about Women* (New York: Crossroad, 1993) 68.

could one exist without the other. On the contrary: they are inextricably linked with one another."[9]

In ancient and medieval thought, the elements were ranked in order of importance from highest to lowest: first fire, then air, water, and earth. Aristotle, who had also designated primary importance to the elements for influencing human constitution and behavior, associated woman with water and earth, the two lowest elements, and man with the two highest elements of air and fire. In contrast, Hildegard "argued that man was more like the highest element, fire, and the lowest element, earth, while woman was more like the two middle elements, air and water. In this way, the two sexes balanced each other out, so that neither one was fully superior or inferior to the other."[10]

As a theological defense for associating the elements with female and male to create a complementary, rather than hierarchical, sex dualism, Hildegard utilized her interpretation of the creation of "Adam and Eve" in Genesis 2. Hildegard wrote:

> Adam, who was created out of the earth, was awakened with the elements and thereby transformed. Eve, however, having emerged from Adam's rib was not transformed. So through the vital powers of the earth, Adam was manly and through the elements he was potent. Eve, however, remained soft in her marrow and she had more of an airy character, a very artistic talent and a precious vitality for the burden of the earth did not press upon her.[11]

Thus, Hildegard associates man with the earth because man was created directly from the earth, while woman is not associated with the earth because she is created from the body of man.

It is important to observe from Hildegard's writings that she does not value one sex over the other, nor does she indicate that humans are separate from or above the elements of nature. Because man is created from the earth, he has power and strength, while woman, who is associated with the air, is artistic and creative. Hildegard's associations are in direct contrast to the associations of women with the earth and men with the cultural refinements of art and creativity that have been prominent in the development of Western philosophy and Christian theology.

---

[9] Hildegard, quoted in Prudence Allen, ed., *The Concept of Woman: The Aristotelian Revolution 750 BC—AD 1250* (Montreal: Eden, 1985) 296.

[10] Ibid.

[11] Ibid., 296–97.

Hildegard's concept of complementary dualism is also evident in her understanding of the relationship between spirit and matter or soul and body, and she uses metaphors from nature to explain this relationship. Hildegard suggests that the soul "wanders everywhere through this form [the body] like a caterpillar spinning silk," and she compares the soul to a great wind: "The spirit of life draws near according to God's will and touches yonder form without the mother noticing, touches it like a strong, warm wind, that sweeps across the plains with a rage; it pours into the form and intertwines into all its limbs."[12]

Hildegard also compares the relationship of soul and body to the "intimate interaction of water in the earth."[13] Hildegard writes: "For as water pours through all the earth, so the soul passes through the whole body," and in another text she declares: "the way the waters dash to particular spots, so the soul infuses our body over which it is all the same superior."[14]

Once again, in contrast to traditional philosophical and theological doctrines, which associated male with the soul and female with the body, Hildegard associates the soul with the same two elements she associates with women: wind (or air) and water. Furthermore, as stated in the last text quoted above, she considers the soul to be superior to the body. However, I hold that this superiority follows the medieval ranking of air and water over earth; Hildegard is not suggesting that women are superior to men.

Hildegard's conception of the equality between women and men is further evidenced in her discussion of the human soul. She believed that because God created male and female in God's image (Genesis 1:27), all human souls had both "masculine" and "feminine" qualities: the female designates mercy, penance, and grace, while the male denotes strength, courage, and justice.[15] The attribution of both feminine and masculine qualities to the soul was not unique to Hildegard, as Aristotle and Philo had posited that the soul included qualities found in both sexes. However, for Aristotle and Philo, the feminine part of the soul was associated with irrationality and lacked authority, while the male part was associated with rationality and authority. Thus, the feminine part of the soul was inferior to the masculine part. In contrast, "Hildegard frequently argued that men

---

[12] Ibid., 301.

[13] Ibid.

[14] Ibid.

[15] Ibid., 298.

ought to develop the feminine qualities of mercy and grace, while women ought to develop the corresponding masculine qualities of courage and strength" as all the qualities were equally important.[16] Although Hildegard did not challenge traditional understandings of masculine and feminine characteristics, she did assign equal value to all these characteristics, and she asserted that both "masculine" and "feminine" characteristics are necessary for a fully developed human being.[17]

In summary, Hildegard associates both woman and man and body and soul with the elements of nature. Her work implies that humans are separate yet integrally connected with nature, as she uses the elements to describe the differences between woman and man and between body and soul. In the pair of woman/man, she offers a theory of sex complementarity that does not value either sex as superior to the other. Furthermore, each sex should seek to develop both the "feminine" and "masculine" qualities of the soul. Hildegard's theory retains the dualisms of man/woman, culture/nature, and spirit/matter, yet she offers understandings of these relationships that are complementary rather than hierarchical; thus, spirit or soul is not separate from nature and neither women nor nature are devalued. There is no justification for domination or exploitation of women and nature in Hildegard's theological anthropology.

Mesoamerican Duality as Fluidity and Balance

When the Spanish invaders came ashore on the land mass now known as Central America or Mesoamerica beginning in 1519, they encountered a "highly developed complex of ideas and beliefs" that constituted the cosmology of the peoples living there.[18] These regional cultures included the Nahua, Mayan, Toltec, Zapotec, and Mixtec.[19] "Despite linguistic differences, the peoples of ancient Mesoamerica . . . . were united by cosmology, mythical and ritual legacy, by cultural centers, and by a similar concept of time and space."[20]

---

[16] Ibid.

[17] Ibid.

[18] Sylvia Marcos, "Cognitive Structures and Medicine," *Curare* 11.2 (1988). The term "Mesoamerica" encompasses the area extending from about half-way between the U.S. border and the Central Mexican highlands, through Mexico and Guatemala, and into parts of El Salvador, Honduras and Nicaragua; Marcos, 87.

[19] Ibid., 87.

[20] Ibid., note 6.

One of the Nahua groups were the Aztecs, also called Mexicas.[21] Sylvia Marcos, who has extensively analyzed Aztec and other ancient Nahua oral traditions,[22] describes the centrality of dualism in Mesoamerican thought:

> Mesoamerican cosmology was based on the idea of dualities and opposites and the search for balance. Duality appeared in every religious duty, political activity and domestic task. . . . Duality in the Mesoamerican cosmovision was not fixed and static, but fluid and constantly changing. This was a key element in Nahua thought."[23]

Thus, both Western and Mesoamerican thought systems conceive of reality as formed by pairs of opposites. However, the Mesoamerican understanding of duality as fluid and constantly changing is in sharp contrast to the traditional Western understanding of the pairs as fixed and mutually exclusive entities.

Marcos also explains that there was no hierarchical valuing of the dualities in Nahua thought:.

> In the various Nahua narratives, whether we look at the *ilamatlatolli* (discourses of the wise old women), [or] the *huehuetlotlli* (speeches of the old men), . . . we can never infer any categorizing of one pole as "superior" to the other. Instead, a sustaining characteristic of this conceptual universe seems to be the unfolding of dualities. . . . Within this fluidity of metaphorical dualities, divine and corporeal, the only essential configuration was the mutual necessity to interconnect and interrelate. In the Mesoamerican universe, above and below did not imply superior and inferior.[24]

"Above and below did not imply superior and inferior." What a novel idea for those of us trained in Western philosophy and theology! The Nahua concept of dualities as unfolding and fluid with the necessity for mutual interconnecting and interrelating provides a basis for defining dualism in a way that does not include the hierarchical valuing of one aspect of the pair over the other.

---

[21] Ibid., note 1.

[22] I was privileged to participate in a workshop with Dr. Marcos in Cuernavaca, Mexico during June 2001. Her presentation regarding Nahua concepts of duality provided the initial spark for my thinking about the possibility of duality as a positive concept for ecofeminist theory.

[23] Ibid., 61, 62.

[24] Ibid., 373.

One example of mutual interconnecting in Nahua thought is found in understandings of how the body is related to the cosmos. Marcos states: "The body's immersion in the cosmos, and the insertion of the cosmos in the body does not allow even the possibility of a body/mind spilt."[25] In contrast to Hildegard, who associated the elements with either women or men, the Nahuas did not assign gender to the parts of the body that were associated with various realms of the cosmos: "The head corresponded to the heavens, the heart as the vital center corresponded to the earth, and the liver to the underworld. These correspondences and interrelations were themselves immersed in a permanent reciprocal movement: the ebb and flow between the universe and the body, and between cosmic duality and the bodies of women and men. . . ."[26] Thus, at the time of the Conquest, the Nahua concept of human existence was more extensive than the developing Christian view of humans as separate from each other and from nature. In contrast to the emerging Christian view, the Nahua perceived humans as forming "a physical and conceptual continuum with others, with the body and with the world beyond it. . . ."[27]

This ancient Nahua concept of a continuum or "permanent reciprocal movement between the universe and the body" has been confirmed by contemporary quantum physics. When humans encounter light, there is an interaction of photons between the light waves and the elementary particles in the human body so that the quantum state of the body is changed. As Brian Swimme explains, "When you stand in the presence of the moon, you become a new creation. The photon's interactions have entered into the quantum state of your entire ensemble, and you are, through these interactions, a moon-person."[28]

Although the Nahuas did not assign gender to the parts of the body associated with specific aspects of the cosmos, the male/female pair did occupy a central position in Nahua cosmology. Marcos writes:

> Deities, people, plants, animals, space, time and the cardinal directions all had a sexual identity as female or male, and this identity shifted constantly along a continuum. . . . Feminine and masculine attributes merged into fluid entities. These expressed the shifting equilibrium of opposite forces which in their turn reflected the

[25] Ibid., 375.

[26] Ibid., 375–76.

[27] Klor de Alva quoted in ibid., 376.

[28] Brian Swimme, *The Universe Is a Green Dragon: A Cosmic Creation Story* (Santa Fe: Bear, 1984) 92.

fundamental balance of the cosmos and society. From the individual to the cosmic, gender appeared as the root metaphor of balance.[29]

These Nahua concepts provide the means for conceiving masculinity and femininity as "fluid entities" that are distinct but not separate. Marcos also describes how fluidity "deepens the scope of bipolarity by giving a permanently shifting nature to feminine and masculine. With fluidity, femininity is always in transit to masculinity and vice-versa."[30] Furthermore, rather than categories denoting superiority or inferiority, the gender designations of feminine and masculine are symbols in the pursuit of balance.

"To maintain this balance is to combine and recombine opposites. This implies not negating the opposite but rather advancing toward it, embracing it in the attempt to find the fluctuating balance."[31] This Nahua concept of embracing the opposite to find the fluctuating balance is similar to Hildegard of Bingen's theory that women should strive to develop the masculine characteristics of strength, courage, and justice, while men should strive to develop the feminine characteristics of mercy, penance, and grace. Both systems of thought develop the concept of complementarity, in which the opposite is desired and attainable, rather than perceived as a separate entity and judged as inferior or superior.

In the Nahua worldview, neither aspect of a pair of opposites is considered superior or inferior because a balance between the two opposites is the goal. According to Marcos, understanding the Nahua concept of balance or equilibrium is necessary for understanding duality from a Nahua perspective:

> Equilibrium determined and modified the concept of duality and was the condition for the preservation of the cosmos. . . . it is a force that constantly modifies the relation between dual and/or opposite pairs. . . . An equilibrium that is always re-establishing its own balance—inherent in the Mesoamerican concept of a universe in movement—also kept all other points of balance equally in constant motion. . . . In a state of permanent movement and continuous readjustment between the poles, neither pole could dominate or prevail over the other except for an instant.[32]

[29] Marcos, *Cognitive Structures*, 62.

[30] Ibid., 373. For a postmodern conception of gender identity as fluid, see Judith Butler, *Gender Trouble: Feminism and the Subversion of Identity*. Rather than shifting along a continuum, Butler argues that gender is constituted by "performance."

[31] Ibid., 374.

[32] Ibid., 373–74.

The Nahua universe is in constant motion, and equilibrium between the opposite pairs is the force necessary to maintain the existence of this universe. Because the poles of the pairs are constantly in motion with each other, sustained domination of one pole over the other is not possible. Thus, in the Nahua conceptual universe, we have a model based on dualism without hierarchy or domination of men over women and nature.

In summary, the Nahua universe of dualities functions without domination due to the Nahua definition of duality. This definition conceives of pairs of opposites that are fluid and constantly seeking a fluctuating balance or equilibrium between the poles of the pair. The force of this equilibrium, as described above in the male/female and human/cosmos pairs, is necessary for the very existence of the universe, a universe characterized by a state of constant motion. This Mesoamerican universe in constant motion is strikingly similar to the conception of reality proposed by Alfred North Whitehead. In the section below, I analyze Whitehead's view of the relationship between body and soul to describe yet another conception of duality that does not promote hierarchy, domination, or exploitation.[33]

Whitehead's Body/Soul Duality of Mutuality

The effectiveness of Whitehead's philosophy in overcoming classical dualism is persuasively argued by Howell. Referencing the work of Mary Daly and Dorothee Soelle as well as her own analysis, Howell describes how Whitehead reinterprets the pairs of Creator/creature or God/world, subject/object, body/mind, and reason/emotion to minimize the "sexist dichotomizing" that has led to the exploitation of women and nature.[34] I build on Howell's argument with my analysis of Whitehead's reinterpretation of the relationship between body and soul.

The association of male with the superior soul and female with the inferior body has been a primary justification for the subordination of females to males. As discussed above in the context of the originality of Hildegard's theory of sex complementarity, these associations from classical Greek philosophy were incorporated into Christian theology and have

---

[33] But were these non-hierarchical Mesoamerican concepts of duality practiced in the relationships between actual women and men? The answer is complicated by the fact that the only written texts of this Mesoamerican cosmology were compiled after the Conquest by Christian missionary friars. These texts do include intriguing indications of gender equity. For example, six men and six women were elected to rule collectively in areas under Aztec control, and careful analysis of Aztec moral precepts reveals many similarities in the admonitions to young women and to young men. See Marcos 1991.

[34] Howell, *A Feminist Cosmology*, 21–23.

functioned to devalue both the body and women. This devaluation has been further exacerbated by the Christian association of sin with women, the body, and sexuality.[35]

An alternative to this traditional Greek and Christian hierarchy of male/soul over female/body is found in Whitehead's ontology of relationality between body and soul. Foundational to Whitehead's conception of how body and soul are related is his principle that reality is composed of occasions of experience. As each occasion of experience is coming into being, it has a "physical pole" through which it receives the influence of the past or actuality, and a "mental pole" through which it entertains the possibility of novelty. Both body and soul are composed of occasions that have physical and mental poles, so a simple association of physical with body and mental with soul is not accurate in Whitehead's philosophy.

The mental pole of every occasion of experience, cellular or psychic, is the occasion's entertainment of new possibilities or novelty. The body is organized to allow one type of occasion, that which constitutes the soul, to be especially affected by novelty and then to transmit that novelty to successive occasions so that the novelty is cumulative. This accumulation constitutes the soul as a "living person." As a result, the soul is the locus of the entertainment of ideas for the whole body.[36] Because the body is organized so that bodily sensations or feelings are "poured" into the ongoing moments of the soul,[37] the occasions of experience that constitute the soul can include more elements of bodily experiences than any other individual part of the body. Thus, the body depends on the soul as a center of organization for the perceptions of the human being.

Yet simultaneously, the soul depends on the body. The body provides the most immediate and most influential environment for the soul as the body mediates the contemporary world to the soul through experiences of space and time. Whitehead explains this bodily mediation as the "withness of the body": "For we feel *with the body*. There may be some further specialization into a particular organ of sensation; but in any case the '*withness' of the body* is an ever present, though elusive, element in our perceptions. . . ."[38]

---

[35] For an extended discussion of this devaluation of women, the body, and sexuality, see Rosemary Radford Ruether, "Misogynism and Virginal Feminism in the Fathers of the Church," 150–83.

[36] Alfred North Whitehead, *Modes of Thought* (New York: Free Press, 1968) 275.

[37] Ibid., 211.

[38] Whitehead, *Process and Reality*, corrected ed., ed. David Ray Griffin and Donald W. Sherburne (New York: Free Press, 1978) 311–12; emphasis original.

Although the soul is not limited to the influence of the bodily environment, for example, the soul can directly prehend the past through memory, the body is primary because the soul is constituted by the actual world of experience mediated through the body. The soul also depends on the body to provide for the ongoing existence of the soul: "The continuity of the soul—so far as concerns consciousness—has to leap gaps in time. We sleep or we are stunned. Yet it is the same person who recovers consciousness. . . . Thus . . . the body in particular provide[s] the stuff for the personal endurance of the soul."[39] Thus, Whitehead understands that the soul depends on the body and the body depends on the soul in a relationship of mutuality.

This process of mutuality through which body and soul develop is also evidenced as body and soul guide each other through shared experiences. Through the body's perceptions of its surroundings, the body guides the soul; the soul receives both the influence of the past and God's best possibility for the person at that moment in the given situation. The soul is free within the context of its present to appropriate the past without being bound to the past. In this way the soul guides the body. The traditional hierarchy of superior soul over inferior body is overcome through Whitehead's metaphysics of reciprocity between body and soul.

Extending his theories beyond the individual human being, Whitehead also describes the relationship between the soul and the world: "The experienced world is one complex factor in the composition of many factors constituting the essences of the soul. . . . in one sense the world is in the soul. But antithetical[ly] . . . our experience of the world involves the exhibition of the soul itself as one of the components within the world."[40] Thus, similar to the Mesoamerican cosmology, Whitehead's metaphysics provide an explanation of the relationality of all creation: the world is a factor in the constitution of the soul, and the soul is one more factor that constitutes the world.

In summary, Whitehead's metaphysics of body/soul and soul/world relationality overturn the traditional hierarchy of superior soul over inferior body and the primacy of humans as above and separate from nature, providing clues for developing an ecofeminist theory of duality without domination. First, the body is not inferior to the soul, because both body and soul are constituted by shared experiences in a relationship of reciprocity as body and soul alternately guide each other. While the body

[39] Whitehead, *Modes of Thought*, 162.
[40] Ibid., 163.

depends on the soul as an organizing center, the soul depends of the body as a source of wisdom for mediating the experiences of the actual world to the soul. Thus, the body should be reverenced, as the body has a primary role in physical and spiritual development.

Second, Whitehead's position undercuts the traditional association of soul/male and body/female, because gender is not assigned to either body or soul. Therefore, the primacy of the body in daily experience is acknowledged for all human beings. Whereas the body is not inferior and woman is not associated with the body, therefore women are not considered to be inferior to men. The justification for women's subordination to men on the basis of women's inferiority is repudiated.

Third, Whitehead's conception of how the soul and the world mutually constitute each other undercuts the traditional hierarchy of humans over nature. Furthermore, in contrast to the traditional association of the female body with nature, Whitehead posits an association of the human soul with the world. The soul is distinct but not separate from the world.

While Whitehead retains the dualisms of body/soul and soul/world, he does not assign gender or superiority/inferiority to either half of the pair. Thus, Whitehead offers a dualistic perspective that resists being used to justify the domination of men over women and nature.

Conclusion

> Radical change in the dominant patriarchal pattern of relationships may require the suggestion of a multiplicity of alternatives to male-defined hierarchy. A variety of concrete options will be necessary for opening the way to real, novel possibilities in human relating.[41]

In response to Howell's call for "alternatives to male-defined hierarchy," I have offered three such alternatives to the classical Western understanding of dualism. These three alternatives support my argument that assumptions of superiority/inferiority, not dualism *per se*, are the basis for the domination of men over women and nature. Furthermore, through a careful analysis of the pairs of man/woman, culture/nature and spirit/matter, I have demonstrated how dualism without hierarchical assumptions can be useful for conceiving the dynamic relationship between entities that are distinct but not separate.

---

[41] Howell, *A Feminist Cosmology*, 14.

The three understandings of dualism without domination discussed above are useful for constructing ecofeminist theory not only because these alternatives dismantle hierarchical valuing of men over women and masculinity over femininity. Each of these understandings also provides clues for how women *and* men are connected to nature through conceptions of body and soul in which the two genders are considered equally valuable, or gender is not associated with either body or soul. Hildegard uses metaphors from nature, Mesoamerican cosmology relates parts of the body to the cosmos, and Whitehead asserts that the soul and the world constitute each other. Positing that both women and men are related to nature provides an answer to one of the remaining "difficult questions" that Howell has identified: "Do women have such a unique connection with nature that culture can never learn the language of communication with nature?"[42] If women and men are equally connected to nature, the answer to this question is "no"; both women and men can communicate with nature and bear responsibility for this communication.

How might ecofeminist theory conceive of such communication between women, men, and nature? Marie Wilson, a spokesperson for the Gitksan-Wet'suwet'en Tribal Council, offers these images of the relationship between males and females and the relationship of humans to the land:

> A North American Indian philosopher has likened the relationship between women and men to the eagle, which soars to unbelievable heights and has tremendous power on two equal wings—one female, one male—carrying the body of life between them. The moment one is fractured or harmed in any way, then that powerful bird is doomed to remain on the earth and cannot reach those heights. . . . I believe that all people started out connected to the land. . . . They saw the cycle of life, from the very smallest to the largest, all connected, . . . They fit themselves into the cycle of life.[43]

The challenge of dismantling hierarchical dualism, so entrenched in Western patterns of thought, is formidable. I hold that conceptions of dualism without domination that describe how all people, women and men,

---

[42] Ibid., 50.

[43] Marie Wilson, "Wings of the Eagle," in *Healing the Wounds: The Promise of Ecofeminism,* ed. Judith Plant (Philadelphia: New Society, 1989) 211, 214 [212–18]. The Gitksan-Wet'suwet'en are one of the First Nations peoples of British Columbia.

are connected to nature can be useful in the effort to end domination and exploitation on this planet.

5

Existence is Relational:
Contemplating Friendship with Nature

Stephanie Kaza

It is my honor to offer further thoughts on Nancy Howell's in-depth intellectual work developing a feminist Whiteheadian perspective. Her thesis contains many points for dialogue and engagement, from the intellectual to the spiritual. Based in both felt experience and well-reasoned thought, the work's authenticity demands equal self-scrutiny and care in applying other perspectives. This paper is one such effort, drawing primarily on Buddhist philosophy, a well as literature in environmental ethics and ecofeminism.

Following traditional feminist protocol, I must acknowledge at the outset my own standpoint as bias. I am not an expert on Whitehead nor a process theologian. I have relatively little meaningful experience either personally or theologically with "God." I am not prepared to take up any of the material in Howell's work that addresses "relating to God." I do, however, bring a strong Buddhist frame to this task, based on twenty-eight years of study and practice in the Soto Zen tradition. My training is both academic and monastic, derived through courses with Masao Abe and others and three years of residence and follow-up training with Green Gulch Zen Center in California. I took lay ordination with Kobun Chino Ottogawa over fifteen years ago and have since studied briefly with several other Zen teachers.

I also bring to this task a well-developed feminist perspective, most familiar with feminist ethics and ecofeminist thought. I have taught ecofeminism for ten years at the University of Vermont and written on the intersection of Buddhist, feminist, and environmental thought. My primary concern, and the focus of all my teaching and writing, is the plight of the planet Earth. As ecological impacts multiply and toxic overload threatens our basic air and water sources, I am compelled to keep my full attention on this global predicament.

Thus, in this essay I will focus primarily on the question of how we relate to nature as human beings. I will work with a selection of Howell's sources to explore further some of the ideas she raises, adding to these sources a range of Buddhist and ecofeminist perspectives. To keep a manageable focus, I will leave to others the equally pressing concerns of how women shall relate to women and men, acknowledging the important contributions Howell has made to this conversation. I see my task here as constructive theology, building on a conversation that has already benefited from considerable dialogue across Howell's choice of materials.

Some Opening Questions

Howell opens her work by asking the question, "why is it important to construct a theory of relations?"[1] She justifies the scope of her project by pointing out the ecological evidence for relationality, the feminist commitment to relationality, and the theological and metaphysical need to provide a relational understanding of God. Likewise, in order to develop this essay, I found myself needing to answer three questions: (1) What is valuable about a feminist cosmology? (2) Why value nature? And (3) Why use a Buddhist lens?

The first question is necessary for me to justify engaging Howell's work at all. Howell makes a strong case for feminist cosmology as a solid alternative to existing frameworks of meaning, based on concepts of God as male, God as controlling power, or God as cosmic moralist.[2] A feminist cosmology provides an opportunity to articulate core critiques of alienation, dualism, and hierarchy in these frameworks.[3] In this, Howell parallels the efforts of ecofeminist authors, particularly those such as Val Plumwood and Karen Warren, who have worked extensively with the problems of stereotyping and dualisms.[4] Such deconstruction is essential

---

[1] Nancy R. Howell, *A Feminist Cosmology: Ecology, Solidarity, and Metaphysics* (Amherst, N.Y.: Humanity, 2000) 8.

[2] Ibid., 29.

[3] Ibid., 21–22, 41–42, 48–50.

[4] See, for example, Val Plumwood, *Feminism and the Mastery of Nature* (New York: Routledge, 1993); and Karen J. Warren, *Ecofeminist Philosophy: A Western Perspective on What*

before exploring constructive theology based in feminist themes such as friendship and relational caring.

Perhaps of equal importance to Howell is the opportunity to develop her own personal spirituality and understanding of God, to undertake "an experiment," an integrative process that reveals new connections.[5] In this undertaking, Howell submits to the arduous journey of meaning-making that is central to the religious quest. While it is comparatively easy to critique others for the shortcomings of their assertions, it is far harder to clarify one's own thoughts as alternatives. Formulating a feminist cosmology provides a challenging spiritual path, fraught with the full range of doubts, rage, surrender, penetration, and acceptance. If one does it completely, she moves from defining a singular personal cosmological view to acknowledging the grand multiplicity of views, some more functional and helpful than others. From both a feminist and Buddhist perspective, this journey is profound and is well served by taking up the project of feminist cosmology.

The second question takes up the primary concern: why value nature? Activists, philosophers, nature writers, biologists, and others have been motivated to identify values of nature, both out of moral responsiveness and out of a desperate need to counter the reigning instrumental values of nature. Valuing nature—whether spiritually, recreationally, or ecologically—seems critical to slow the accelerating destruction of global forests, waterways, oceans, and atmosphere. Environmental ethicists such as Holmes Rolston III and Warwick Fox have cataloged lengthy taxonomies of reasons to value nature.[6] Howell reiterates the compelling need to lift up other human values in relation to nature besides those dualistic, hierarchical, and anthropocentric views that reinforce the superiority and complexity of humans.[7]

More fundamentally, valuing nature is valuing human existence, since humans are utterly dependent on the matrix of nature for life support. Writing about nature and value becomes a way to point to the pre-existing web of relations across time and space that gave birth to human existence.

It Is and Why It Matters, Studies in Social, Political, and Legal Philosophy (Lanham, Md.: Roman and Littlefield, 2000).

[5] Howell, *A Feminist Cosmology,* 14–15.

[6] See Holmes Rolston III, "Values in Nature," in *Philosophy Gone Wild: Essays in Environmental Ethics* (Buffalo, N.Y.: Prometheus, 1986) 74–90, and Warwick Fox, *Toward a Transpersonal Ecology: Developing New Foundations for Environmentalism* (Boston: Shambhala, 1990).

[7] Howell, *A Feminist Cosmology,* 37.

Feminist values in relationship to nature, such as kinship, mutuality, subjectivity, and diversity, are now being adopted by advocates for ecological sustainability. Howell points to two Whiteheadian values of intensity and potentiality as core to a relational understanding of nature.[8] She raises the intriguing possibility of extending these values from the individual experience to the community experience. Examples of this vision can be found in communal indigenous relations with nature, as well as certain alternative communal religious sects.[9]

In short, the project of valuing nature may be essential to the continued existence of humans on the planet. It is far more than an academic exercise. The urgency behind Howell's work is shared by many. As conversations such as these reach further into society, there is some hope for turning the tide of unsustainable practices based on inadequate values. In this sense, every effort counts, even though from the small scale of a human lifetime it is difficult to predict the outcome of such experiments.

Why engage a Buddhist lens? Interfaith dialogue, in general, can be helpful in highlighting the commonalties and differences between religious points of view. Buddhism, in particular, has drawn extensive interest from those who perceive it as "wholly other" to western monotheistic faiths. Over the past fifty years, since the earliest explorations by Thomas Merton and others, Buddhist-Christian dialogue has accumulated a significant history in the academic community as well as in faith-based churches and monastic traditions.[10] At this point, extensive discussion in formal and informal settings has generated considerable mutual understanding across religious difference.

Process theologians such as John Cobb have been instrumental in encouraging this dialogue, sensing many points of overlap with Buddhist philosophy. Both approaches support a foundational view of reality as dynamic and ever-changing, always in process even within material form. Both try to analyze the units of experience and the stages of becoming—what brings something into being, how an event unfolds. In contrast to some absolutist Christian perspectives, process theology has a more flexible, open understanding of God, making dialogue with a nontheistic tradition more possible. As for feminist concerns, Buddhism, like Whiteheadian

---

[8] Ibid., 57–60.

[9] Bronislaw Szerszynski, "The Varieties of Ecological Piety," *Worldviews: Environment, Culture, Religion* 1 (1997) 37–56.

[10] See for example, Susan Walker, ed., *Speaking of Silence* (New York: Paulist, 1987); and Bhikshuni Thubten Chodron, *Interfaith Insights* (New Delhi: Timeless, 2000).

thought, does not address these adequately. Yet also like process theology, Buddhism can provide support for a feminist perspective.[11]

Perhaps the most important and relevant aspect of Buddhism with regards to this essay is its fundamental insight as to the relationality of existence. The law of mutual causality or dependent co-arising has been central to a Buddhist worldview across all Buddhist traditions from the first teachings of the Buddha. This law was particularly emphasized during the Hua-Yen period in China, where key texts described the world as Indra's Net, a vast interconnected web of infinite dimensions, with shining jewels at every node, each jewel with infinite facets reflecting all the other jewels in the net. The Buddhist religious heritage offers a rich wealth of teachings on relationality, studied in great depth and philosophical detail, with extensive implications for how we live ethically with and within nature.

Buddhist Views of Relationality

In order to apply a Buddhist view to Nancy Howell's work, it will be necessary to examine key aspects of the Buddhist sense of relationality and see how these compare with a Whiteheadian view, as described by Howell. Following this exploration, I will look more closely at two particular aspects of relationality that are crucial to Howell's feminist cosmology. These are relationality with *nature* and relationality with *power*. We will see what Buddhism can offer in engaging these significant and penetrating aspects of human relationship.

Existence is relational: what does this mean? The Buddhist law of mutual causality states that all phenomena exist *inter*dependently, that no phenomenon exists *in*dependently. All beings, events, and material manifestations reflect multiple causes and conditions that brought them into being. In this sense, the phenomena are dependent on these causes and conditions for their nature. A windy headland will produce windblown pines and firs. A toxic lake will produce developmentally deformed frogs. Endless pounding of rain on rock will generate erosion, landslides, and muddy rivers. Likewise, human beings reflect parental values, household conditions, community norms, local climate, and many other large and small-scale influences on each life. The corollary to this truth is equally important: that no being can be said to have a separate self since each self is composed of many elements.[12] The western notion of a "self-made man"

---

[11] See for example, Rita M. Gross, *Buddhism After Patriarchy: A Feminist History, Analysis, and Reconstruction of Buddhism* (Albany: State University of New York Press, 1993).

[12] Vietnamese Zen teacher Thich Nhat Hanh has developed this extensively with his

is anathema to Buddhism, since such a thing cannot truly exist. Rather the man is made of his culture, his peers, his tasks, his attitudes, his landscape. "Self," then, as traditionally understood in western philosophy, is a problematic term from a Buddhist perspective. Many Buddhist meditation techniques have been developed specifically to illuminate the human tendency to grasp after a false view of self as permanent and unchanging.[13]

Causality is likewise understood quite differently from a Buddhist perspective. Whereas western traditions emphasize linear single cause explanations for phenomena, Buddhist philosophies emphasize multiple causes, asserting that a single cause is seldom adequate. Causality can and does act on any scale, from macro to micro. Much of what humans are capable of observing about causality may only be the proverbial tip of the iceberg. We might, for example, attribute a particular diseased condition, say having a cold, to being exposed to someone with a cold. But other contributing causes may include climate fluctuation, immune system degeneration, emotional stress, etc. We can speak of this as the scale and degree of *mutual interpenetration*.[14] To what extent do particular causes influence particular phenomena? This is the life work of geologists, ecologists, soil scientists, as well as health professionals. But it is also central to all Buddhist analysis of relationality. Other philosophers speak of *mutual responsiveness*, a sense of endlessly interweaving conversation between all phenomena.[15] I will return to both of these aspects when we look at nature and power from a Buddhist relational view.

To illustrate the dynamic interdependent nature of reality, let us look at what all beings hold in common: arising and passing away. Buddhists describe impermanence as a key characteristic of all existence. Let us see how such "coming into being" and "passing out of being" is relational in nature. We could choose anything, but perhaps imagine a particular oak tree. Its coming into being depends on geophysical conditions, bioecological relations, and some version of social relations as well. As an acorn seed sprouts it encounters air and soil conditions—are they favorable? or will it be dealing with erosion, wind, depleted nutrients? Once the hull cracks open, the nutritious meat or first leaves are easy food for mice, deer, insects. As a young oak coming into being, the seed enters a world of bio-

phrase "interbeing"; see, for example, "The Sun My Heart," in *Love in Action*, 127–38.

[13] Zen koans in the Chinese and Japanese traditions are particularly famous for their teaching effectiveness.

[14] Thich Nhat Hanh, *Love in Action*, 134.

[15] For a Buddhist perspective, see Honghzi, *Cultivating the Empty Field* 1991. For a western perspective, see Abram, *The Spell of the Sensuous*.

logical relationships that will determine its fate. Social relations among the local plants and animals will affect who inhabits the immediate neighborhood, how safe it is, whether or not it will be developed or protected by humans. The seed has no choice but to be deeply influenced by this world of relationships.

Likewise, a phenomenon cannot pass out of being independently; the cessation of being is also relational. Geophysical conditions stretch a body beyond what it can tolerate: the branch snaps in a gale, starfish bake in an overexposed tide, sled dogs freeze in subarctic wind chills. Ecological relations may bring death by predation or parasitism, competition or infection. Changes in social relations—abandonment, neglect, or the opposite, support and encouragement in dying—carry tremendous influence on how a being ceases to exist. Passing away does not happen outside of relationality but is the very manifestation of that relationality.

One of the best-known attempts to communicate a relational worldview was undertaken by Fa-tsang, a Chinese patriarch of the seventh century.[16] As a display for the Empress Wu, Fa-tsang had an entire palace hall lined with mirrors. At the center he placed a torch and a Buddha figure. The spectacular vision of the Buddha reflecting endlessly through the mirrors was designed to give the empress a glimpse of interpenetration. The myriad Buddhas represented various phenomena across space, though, as Fa-tsang explained, the metaphor did not capture interpenetration across time. The Net of Indra offered another metaphor, contributing further understanding of relationality. A tug on the net or cloudiness in a jewel affected others in the web. Likewise, the polishing of virtue by any single being brought greater clarity and capacity to the web of relationships. Thus, the teaching of interpenetration became an avenue for promoting ethical behavior, always seen in a relational context.

The Net of Indra has been widely adopted by Buddhist environmentalists as a useful expression of ecological relationship.[17] Yet as a metaphor it has its limitations. For one thing, it indicates a false sense of uniformity; it does not describe the specific nature of individual relationships. For another, the web metaphor implies an equality of relations, as if all phenomena are equally influenced by all others. This overlooks the scale and degree aspect of influence. A third criticism is that the metaphor tends to be interpreted spatially, not temporally, emphasizing existing relation-

---

[16] Described in Heinrich Dumoulin, *Zen Buddhism: A History*, trans. James W. Heisig and Paul Knitter (New York: Macmillan, 1988) 1:47.

[17] See, for example, selections in Stephanie Kaza and Kenneth Kraft, editors, *Dharma Rain: Sources of Buddhist Environmentalism* (Boston: Shambhala, 2000).

ships over those generated across time.[18] Thus it is important to remember Indra's Net as metaphor, not complete teaching. The deeper understanding of relationality comes from penetrating its full measure from within the experience of it. The teaching of the Ten Interpenetrations gives some inkling of this profound perspective. For example, "One perception penetrates all perceptions. All perceptions penetrate one perception." Or "All times penetrate one time. One time penetrates all time—past, present, and future."[19]

With this introduction to Buddhist relationality, let us now compare the Buddhist view with the Whiteheadian view described by Howell. Acknowledging my own lack of training in process theology, I take this material directly from Howell's thesis. I will look at three concepts she relies on: becoming ontology, internal relations, and causal efficacy. The Whiteheadian *ontology of becoming* (vs. being) gives priority to process over substance.[20] This is central to a process view of the world as dynamic and ever-changing. Buddhists would confirm this dynamic view, but would go even further, seeing all process *and* all form as continuous flux. The Heart Sutra chanted daily in many Zen monasteries, states unequivocally, "Form is emptiness, emptiness form," emphasizing the changeability of all phenomena.[21]

Likewise, paralleling the Whiteheadian process of "concrescence," Buddhist philosophy has also taken up the challenge of understanding single moments as manifestations of potentiality. But Buddhists would be just as interested in the coming apart or disintegration of these moments as well as their creation. This perhaps points to a basic difference between process theology and Buddhist philosophy. Buddhism emphasizes the study of death or the cessation of being as a more useful teaching than the act of creation. This is because human beings will naturally start from an assumption of existence (i.e., creation) based on their own experience as having come into being. To study death or dissolution is a more useful corrective, helping students see their lives as impermanent and interdependent phenomena.

---

[18] Ian Harris, "Buddhism and Ecology," in *Contemporary Buddhist Ethics*, ed. Damien Keown (Surrey, Eng.: Curzon, 2000) 113–35.

[19] Thich Nhat Hanh, "The Sun in My Heart," in *Love and Action: Writings on Nonviolent Social Change* (Berkeley: Parallax, 1993) 134.

[20] Howell, *A Feminist Cosmology*, 23.

[21] See Masao Abe, *Zen and Western Thought*, ed. William R. LaFleur (Honolulu: University of Hawaii Press, 1985) for commentary on this sutra and the concept of emptiness.

Does this difference reflect the Christian emphasis on a creator God? There is no counterpart for such a god in Buddhism, since it is a nontheistic religion. For Buddhists, the basic human problem is clinging to a false view of the self as permanent. *Nothing* is permanent in the Buddhist dynamic worldview. This presents a very distinct difference from the Christian understanding of God as absolute and unchanging in contrast to all other phenomena. "Inevitability of death" vs. "God as creator" pose quite different entrance points for arriving at a view of existence as dynamic process. Emphasis on "coming into being" to the point of ignoring or overlooking "passing out of being" can lead to a distorted view, missing key aspects of relationality.

Howell relies heavily on Whitehead's doctrine of *internal relations* to frame her feminist cosmology.[22] "External relations" seem to represent an *in*dependent view of existence, as described by Howell, whereas "internal relations" represent an *inter*dependent view, with multiple causes and conditions influencing any single being. Reframed in these terms, the overlap with Buddhism is strong, with Howell choosing internal relations as core to her cosmology and experience. But the worldviews diverge with the introduction of "freedom" and the "divine persuasive lure," For Whitehead, "freedom" indicates a certain degree of self-control, initiative, creativity—an apparent characteristic of the initiating being that determines the nature of relationality. This would conflict with a Buddhist emphasis on "no separate self," i.e., nothing is independently determined. Freedom, as used here, does not have much credence within Buddhist philosophy. I appreciate Bev Harrison's reshaping of this concept as participating with others in co-creating the world. This is much closer to the Buddhist view of agency within the interdependent web. "Divine persuasive lure," however, does not, as far as I know, have a Buddhist counterpart. Being nontheistic, Buddhism has no sense of an all-pervasive creative force or presence that beckons. Such a force seems to me to be very distinctly Christian or at least tied to theistic views. Internal relations for Buddhists, in context, would be composed only of elements in relationship, with no larger force beckoning or calling into being. I like Catherine Keller's characterizing of this as "in-fluence," "not working *upon* me so much as *into* me; influence is that which flows in."[23]

Causal efficacy, as a third aspect of Whiteheadian theology, is to the best of my understanding, a mode of perception of the present that ac-

---

[22] Howell, *A Feminist Cosmology,* 25–26.

[23] As cited in Howell, *A Feminist Cosmology,* 91.

knowledges influences from the past and determines how much these will affect the emerging subject.[24] This sounds as if it is self-reflective and self-determinative, a thought process initiated by an engaged human being. What would be the counterpart for other-than-humans? Does this concept work as well for animals, plants, rivers, mountains? Though likely not the same, I am reminded of Buddhist views of cause and effect, otherwise known as *karma*. Here causation is much more action driven than reflective. Causes of actions include behaviors, thoughts, and intentions; consequences of actions reflect all the multiple causes and conditions that shaped them. This seems to me to be a much more all-inclusive concept of causal relationality.

As one last piece of this foundational dialogue, I want to compare process theology and Buddhist views of the role of experience in perceiving relationality, as this is central to both philosophies. For Whitehead (as explained by Howell), the experiencing subject is the locus of understanding; "experience precedes theory."[25] This emphasis represents a significant corrective to other western non-embodied abstract and generalized interpretations of the world (problematic for feminists for similar reasons). Howell cites Whitehead's catalogue of many types of experience to which he urges us to be receptive. This "new logic" based on experience eschews universalizing, the philosophical tendency to generalize about an insight across different cultural and individual experiences, a trend well-noted by feminists as typical of patriarchally-based worldviews. But I believe that Buddhism takes this focus on experience even farther, providing complex methods for reporting and evaluating individual experience.

From a Buddhist perspective, personal experience *is* the locus for enlightenment. Awakening and liberation can *only* happen in the realm of experience; it is not an abstract thought event. The Buddha was adamant that followers should test his teachings for themselves, personally engaging in the pursuit of authenticity. Yet he also pointed out that experience itself is not significant per se, since it is ongoing. The more critical challenge is to wake up in the middle of that experience and to see that it is empty of independent existence. In this sense, Buddhism does not merely acknowledge the importance of experience, as Howell does, but it posits the entire project of liberation on the necessity of experience.

---

[24] Ibid., 92.
[25] Ibid., 16.

Relating to Nature, Relating to Power

Having laid out a foundation for understanding Buddhist views of relationality, we now come to the applied part of this paper. How can Buddhist
views shed light on the way people relate to nature and to power? These
are among Howell's primary concerns, as she is highly motivated to gain
insight into human-nature relationships in all their multiple destructive
and constructive forms.

Though work in the field of Buddhism and ecology or nature is fairly
recent and as yet hardly comprehensive, I can suggest three types of human-nature relationships of importance in Buddhist teachings. First, nature is often seen as *teacher*, offering wisdom and insight to students of
nature. This metaphor carries great weight in Asian culture, as spiritual
teachers are revered and appreciated with great gratitude. Seeing nature
as teacher is a way to acknowledge its tremendous capacity for serious
learners. In the Lotus Sutra, metaphors of abundance and proportionality
are used to indicate how teachings flow like rain in accordance with need.
Elements of nature are often used to model qualities for human virtue: the
endurance and patience of pine, the flexibility of willow, the flow of water.
In the Mountains and Waters Sutra, 13th century Zen teacher Dogen
exhorts students: "Wherever there is a world of sentient beings, there is a
world of Buddha ancestors. You should thoroughly examine the meaning
of this."[26]

Nature is also seen as *refuge*, a relationship of protection for practice.
Milarepa's famous cave is one of many Tibetan cave refuges that support
multi-year solitary meditation retreats. Japanese poet monks speak of their
small mountain huts or taking solace under pine trees. Forest-dwelling
monks of southeast Asian Buddhism still use the lush trees of the rainfor-
est as shelter during the rainy season when they stop their wanderings.
Often this sense of nature as refuge refers to a broader sense of nature as
original home; returning to nature is returning home to one's own original
nature.

In much of the Chinese literature, nature is seen as imbued with
Buddha-nature much as humans are. Buddha-nature is seen as pervasive,
universal, accessible, and full of potential to be expressed in myriad forms.
In Japanese texts, the emphasis shifts to a nondualistic view, seeing *no
separation* between nature as other and nature as self. This is not some
blurring of distinctions into a single unity of experience, but is rather an
understanding of the complex and simultaneous interdependence between

---

[26] For this sutra and other texts involving nature, see Kaza and Kraft, *Dharma Rain*.

experiencing elements of nature, including humans. This is, then, not a specific type of relationship as in "teacher" or "refuge" but rather the total experience of endless relating as context for all activity.

As a side note, I should add that there is not too much in the Buddhist literature to indicate significant gender differences in relating to nature. There are relatively few women writers across the long history of Buddhist philosophy, for as a monastic tradition, it was more often monks who set the Buddha's teachings in text form. Thus, there are few resources here for an ecofeminist Buddhism, though a few modern writers, including me, have attempted some preliminary explorations.[27]

What does Buddhism have to offer in the study of relationships of *power*? This is somewhat less obvious than relationships with nature. Buddhism is concerned primarily with spiritual development, so abuse or distortion of power would be seen first as a problem for the afflicted individual. Such non-virtuous behavior would be diagnosed as a manifestation of the three poisons of greed, hate, and delusion. The antidote is then to study and penetrate the nature of these universal poisons as they appear in one's self. For this, the Buddhist texts provide excellent detailed techniques for studying the emotions that drive power relations, such as fear, anger, and confusion.[28] Emotions are not seen as secondary to reason, as in much of Western philosophy, but are instead the subject of intense scrutiny. In-depth understanding of one's wide-ranging emotional experience can bring stability and insight to power abuse situations.

Because Buddhism focuses directly on suffering and the relief of suffering as the framework for liberation, it has the capacity to analyze power relations through this lens. Who is suffering and from what? How can their suffering be relieved, whether personal or structural? This kind of diagnosis is being developed by modern socially-engaged Buddhists to shed light on particular institutional forms of power abuse, as in the military or government.[29] Buddhists explain quite easily how desire and attachment to privilege, possessions, status, and power inevitably cause suffering for

[27] See Gross, *Buddhism After Patriarchy*; and Kaza, "Acting with Compassion: Buddhism, Feminism, and the Environmental Crisis," in *Ecofeminism and the Sacred*, ed. Carol Adams (New York: Continuum, 1993) 50–69.

[28] See for example, Herbert V. Guenther and Leslie S. Kawamura, trans., *Mind in Buddhist Psychology*; and Kalupahana, *The Principles of Buddhist Psychology*, SUNY Series in Buddhist Studies (Albany: State University of New York Press, 1987).

[29] See Sulak Sivaraksa, *Seeds of Peace: A Buddhist Vision for Renewing Society*, ed. Tom Ginsburg (Berkeley: Parallax, 1992); and idem, *Santi Pracha Dhamma* (Bangkok: Santi Pracha Dhamma Institute, 2001).

others. The Mahayana Buddhist goal to relieve others of suffering is expressed as the Bodhisattva vow, to return again and again to save *all* beings from whatever extreme of suffering they endure. The power of this vow can be seen as a mandate for engaging social justice concerns, and members of groups such as the Buddhist Peace Fellowship are taking these up with deep practice motivation.

However, in keeping with Howell's rigorous scrutiny, I must acknowledge here the well-described shortcomings of the Buddhist approach to dealing with power. In contrast to both feminism and process theology, there is no comparable project to examine power, particularly hierarchical power, and its meaning. Howell explores a "dipolar thesis" of power, Cobb investigates "persuasive power," Harrison discusses "reciprocity in relation"—where is the counterpart in Buddhism? I wonder if perhaps Christian conversations on power may be more evolved because, as part of their theological inquiry, Christians must confront the nature and role of an "all-powerful" God. How does a small and vulnerable person relate to such an awesome being/force? Is God's power the power to command, to love, to forgive, to endorse? Is it kind, cruel, compassionate, indifferent? With no comparable theological problem embodying the "p" word, Buddhists are perhaps less driven to engage these questions.

Instead, more attention has been paid to *karma* and its many implications. Often one's status or fate is attributed to the experience of a past life influencing one's current incarnation. This simplistic individualistic view has been critiqued for generating a passivist approach to political and economic conditions. Narrow interpretations of karma place the blame for any experience of suffering directly on the person who is suffering. This overlooks widespread socially embedded racism, class privilege, gender bias, political oppression, etc. The lack of Buddhist structural analysis regarding power is frustrating for those of us concerned with suffering from power abuse, leaving us relatively little historical guidance to draw on.

As for relating to power, Buddhist peace activists promote a nondualistic approach: seeing the enemy as co-creator of a situation of suffering, rather than separate and immovable. Demonizing the enemy does not work from a nondualistic view; it is too polarized and does not represent the truly influential nature of interdependent relations. But seeing the other and self as co-created rather than oppositional is a very difficult practice, particularly when one half of the relationship is causing tremendous suffering to the other half. The simplistic good/evil dichotomy pervasive in Western philosophy and much of Christian theology is not very useful from a nondualistic Buddhist perspective. Ecofeminist philosopher Karen

Warren has laid out a logic of domination in which superiority justifies subordination.[30] She shows how this logic is reproduced in racism, sexism, classism, and "naturism"; in each pattern of oppression, oppositional dualisms evoke the related dualisms, reinforcing up-down relationships of power. Buddhist nondualism offers an antidote to this all-pervasive social patterning, but, admittedly, it is not a significant counterforce in today's troubled world.

Friendship, Nature, and Power

Exploring relationality can perhaps best be handled by looking at specific kinds of relationships. Following Howell's interest in *friendship* as one category of relationship, I want to look here at aspects of friendship relevant to nature and power, our focus above. But how shall we understand the term "friendship"? I will begin by reviewing five aspects of friendship Howell has chosen to emphasize and offering a Buddhist response to these. Then I will consider how these aspects may be relevant in framing friendships in relation to nature and friendships in relation to power.

Sidestepping for the moment the merits of separatism, I want to go directly to where this and other viable avenues lead one in understanding friendship. First, Howell suggests true friendship has the capability of offering *liberation* from false views of self. In this it holds much in common with most religious traditions, though Howell focuses more specifically on paring away "the false selves layered upon women's selfhood by patriarchy."[31] From a Buddhist view, this process of liberation from false views of self is central to awakening, but it is applied very broadly. *Any* constricting misperceptions can be points of self or ego-construction and are thus ripe for insight into interdependence. The Buddhist concept of spiritual friendship or *kalyāṇamitra* suggests a type of friendship based in the work of liberation, i.e., helping one another break through ego-bound forms of self.

Corollary to liberation from false patriarchal views is Howell's concept of *empowerment* or development of [women's] selfhood. Here one is called to become a friend to one's self, or one's own "be-ing," redirecting energy away from the maintenance of patriarchy and patriarchal views of self and toward greater personal integrity. For women, Howell suggests, this is a pro-active process, necessary for recovering what has been lost or buried under patriarchal culture. In southeast Asian Buddhism,

---

[30] Warren, *Ecofeminist Philosophy,* 48.

[31] Howell, *A Feminist Cosmology,* 69.

this sense of befriending oneself is addressed in the Metta Sutta or loving kindness meditation.[32] First one offers loving kindness to oneself, then to family, then to friends, enemies, and eventually to all beings. Starting with one's self is critical to establishing the practice within before extending it outward. In Zen, however, this process of self-love or development seems perhaps more neglected. Much of historical Zen training in China and Japan was directed at male students, apparently in need of significant ego-busting. When Zen arrived in the west, it encountered many more women students and with them, significant issues of low self-esteem. It has taken some dialogue and sustained effort to find ways to encourage self-befriending in the modern Zen context.

A third feature Howell brings out is friendship as *co-creative*, a shared journey engaging both self and other. As a more liberated, creative self emerges through the process of Be-Friending (described by Daly), it beckons to others to join the process and become likewise liberated. Daly uses the term "en-couragement," to be filled with courage, so much so that the creativity of that courage becomes contagious, forming a multiplier effect for women's friendships.[33] A Buddhist counterpart of such sisterhood would be the *samgha*, or community of practitioners following the Buddha's way. In a way similar to Daly's sense of friendship, Buddhist students cultivate self-knowledge through interacting and learning with others. Emphasizing the nature of this co-created journey, Buddhist feminist Rita Gross has developed an extensive interpretation of *samgha* as the necessary matrix for spiritual development.[34]

Howell draws on Raymond's philosophy, highlighting her call to *affection for the "original woman."* This is not a romantic essentialist sense of original or an imaginary prototypical self but, rather, a hard-earned self-created self, formed in the process crucible of breaking through oppressive views. "Affection" indicates a deep appreciation for the effort it has taken for someone to become "original."[35] In contrast, a Buddhist view of "original self" is that which connects with the unbroken stream of mind across time. This is qualitatively different from the sense in which Howell or Raymond means "original." From a Buddhist perspective, even the hard-

---

[32] For a modern discussion of this practice, see Sharon Salzberg, *Loving Kindness: The Revolutionary Art of Happiness* (Boston: Shambhala, 1965).

[33] Howell, *A Feminist Cosmology*, 70–72.

[34] Rita M. Gross, *Buddhism after Patriarchy*, 258–69.

[35] Howell, *A Feminist Cosmology*, 73.

earned self-created self may be laden with ego constriction and can be pared away for an even more liberated sense of the emptiness of "self."[36]

A fifth feature of friendship Howell presents is *solidarity*. In Howell's use of the term, she means not only supporting others in the struggle for liberation, but embracing differences with others, i.e., being able to accept and tolerate and also stand behind people whose process or values are different than yours. This, she feels, is crucial for sustaining and nurturing the overall women's project of liberation. If liberation were only an individual process, it would eventually die out, and widespread transformation of oppressive institutional structures would not be possible.[37] Embracing differences requires not only self and other friendship, but self-critique and accountability. I am not aware of a counterpart Buddhist term or value other than *samgha*, but there are modern day examples of Buddhist solidarity in action, such as anti-war activism, the alternatives to consumerism movement, and the self-help Sarvodaya movement in Sri Lanka.

Friendship and Nature

The idea of friendship with nature has been explored in various ways by western nature writers and environmental philosophers. In this constructive part of the essay, I want to see if Whitehead process theology can be similarly applied to friendships with nature, or more specifically, beings in nature. Certainly, for starters, the Whiteheadian view of *continuity* between humans and nature is supportive of some form of friendship. Ecofeminists and Buddhists would both support this sense of continuity, though degree of continuity varies somewhat depending on the particular Buddhist tradition.

Whitehead's definition of "*person*" is also relevant to this question, as he includes animals as persons, saying they have "an internal organizer that acts as a social coordinator of the society of occasions that make up a living body."[38] To me, however, this application of person to animals does not seem biologically grounded, for it appears to assume that shared neurological or behavioral qualities are common to all animals. Considering the limited capacities of such organisms as sponges or corals, one sees that this level of coordination varies considerably. And why not plants as persons then? There is evidence that similar coordinated organization takes

---

[36] See for example, Dogen, "Genjo Koan, Actualizing the Fundamental Point," in *Moon in a Dewdrop*, ed. Kazuki Tanahashi (San Francisco: North Point Press, 1985).

[37] Howell, *A Feminist Cosmology*, 81–85.

[38] Ibid., 28.

place in plants, though it is chemically rather than neurologically based. From a Buddhist view, it is more important to focus on the commonalties of impermanence and suffering than the categories of coordination and experience. Still, there is debate between Buddhist lineages over whether or not plants experience suffering.[39]

A more significant problem in developing a concept of friendship with nature is Whitehead's hierarchical categories of the four aggregates. Here I support Howell's ecofeminist critique, affirming that Whitehead's view is too reflective of human projections and not ecologically accurate. ("Taxonomic hierarchies do reflect degrees of relationality, but these are indicators of evolutionary history, not necessarily ecological relationship.") Other religions and ethicists have wrestled with assigning gradations of value to members of the natural world. The Jains, for example, rank beings according to the number of senses with which they perceive the world. Though Howell tries to work with Birch and Cobb's interpretation or modification of Whitehead's four aggregates through a definition of intrinsic value, I find that the hierarchy remains intact. Balancing instrumental value with intrinsic value still presents a hierarchical ranking of organisms, with degrees of moral permission, e.g., it is permissible for humans to use rocks but not animals (based on degree of intrinsic value). Giving greatest value to those beings most capable of "rich" experience can justify choosing for these over other beings less capable. From a Buddhist perspective, this ignores a much broader interdependent view and the infinitely complex range of relationships that both support and threaten humans, richly or not. The categories of intensity and lure of God do not seem relevant from a Buddhist view.

Setting aside this critique for the moment, let me turn to the highlighted aspects of friendship and ask, can these apply to friendships with specific beings in nature? The first two, *liberation* and *empowerment,* do not seem to apply directly, as these are processes particular to the human predicament. However, I believe that engaging in these would increase one's capacity for forming friendships with other-than-human beings. The third, *co-creative self-knowledge* through others seems more promising; one can take seriously the possibility of meeting trees, rivers, or mountains through deep attention and receptivity. Writer Terry Tempest Williams, for example, expresses her deep friendship with the desert canyonlands of the Southwest.[40] Buddhist poet Gary Snyder speaks of reinhabitation

---

[39] LaFleur, 136–44.

[40] See her most recent book, Terry Tempest Williams, *Red: Passion and Patience in the*

with Coyote, Deer, Jay, and the other friends of his home area in the Sierra foothills of California.[41] Among environmental educators there has been a virtual explosion of interest in place-based education with the explicit goal of cultivating friendships with rivers, trees, and birds.

How about *affection for "original nature"*? To apply this concept to beings in nature, one would need to appreciate the evolutionary developmental journey behind any natural form. This is at the heart of Thomas Berry's ecotheology, expressed in the Universe Story, the wider human narrative he suggests could be shared by all religions (Swimme and Berry). Raymond's concept would transfer well in cases of restoration of damaged areas, where human engagement with natural processes has rebuilt the health of a beloved friend (farmland, forest, wetland, etc.). Here affection develops through the work of paring away the damage and providing supportive measures for recovery. Damaged nature becomes "original" through dedicated effort over time—pulling invasives, planting trees, reintroducing predators.

Of the five aspects, I believe *solidarity* may have the greatest potential for building friendships with beings in nature. Embracing differences across human and other-than-human beings is prerequisite for solidarity with nature and far from easy. Some differences are easily revered (the sonar of dolphins, the strength of oaks) while others are disparaged (the sting of wasps, the tenaciousness of ivy). Some may find it possible to feel solidarity with only certain members of nature. But from a Buddhist perspective, this would not be going far enough. Solidarity as expressed in the Bodhisattva vow would mean taking a stance on behalf of all beings facing oppression or exploitation, supporting their struggle for existence as a call of friendship. The now-famous Chipko women of India took this position on behalf of their forests, which were threatened by government logging. Hugging the trees, they stood in their defense and protected their friends from attack.

In addition to the aspects of friendship Howell outlines, I would like to add several others that may be useful in developing friendships with beings in nature. Karen Warren suggests regarding other beings with "the loving eye" or what she terms "caring perception."[42] She contrasts this to "the arrogant eye" that looks to nature for conquest rather than intimacy.

Desert (New York: Pantheon, 2001).

[41] See his essays in *The Practice of the Wild: Essays* (San Francisco: North Point, 1990); and *A Place in Space: Ethics, Aesthetics, and Watersheds—New and Selected Prose* (Washington, D.C.: Counterpoint, 1995).

[42] Warren, *Ecofeminist Philosophy*, 104–5.

The loving eye pays close attention as it encounters the other, acknowledging both the relationship that is forming and the distinct, independent complexity of the other being. She suggests a feminist narrative method for cultivating friendships with rock, maple, or raccoon, based on sharing stories of time spent together. The stories help shape the bond of the friendship, even though they are told only through the human form.

From the Buddhist tradition, I can offer two bases for friendship with nature. In the earliest Jataka Tales of the Pali Canon, the Buddha is reborn across many lifetimes as a compassionate tiger, rabbit, tree, etc. The stories convey an underlying message of kinship with all nature through the process of rebirth.[43] One is urged to be friendly with all beings in nature because they may be a rebirthed form of a loved one. This kinship can be seen today in modern protests against logging, when local Thai monks ordain large village trees, affirming their longtime friendship with particular venerables.[44] East Asian traditions hold the belief that plants and animals are also destined for Buddhahood. In Japanese Noh theater, plants, particularly trees, often have leading roles, and it is unclear whether their journey in the play is metaphoric for humans or directly pointing to the plants attaining of Buddhahood.[45] Zen master Dogen speaks of "mutual attainment of the way," indicating the path of liberation is simultaneously realized by all beings.

Friendship and Power

In what way can the relationship of friendship be useful in addressing power? To answer this, we must first articulate some definitions of power. Howell links destructive use of power with patriarchal oppression, constructive use of power with women's liberation from that oppression. She defines power as the "ability to make choices."[46] I would add to this a broader expression used by Warren, that power is the "ability to mobilize resources to achieve desired ends."[47] Warren lists five kinds of power: (1) power over, as in domination and oppression; (2) power with, as in actions

---

[43] Harris, "Buddhism and Ecology," 119.

[44] Susan M. Darlington, "Tree Ordination in Thailand," in *Dharma Rain: Sources of Buddhist Environmentalism*, ed. Stephanie Kaza and Kenneth Kraft (Boston: Shambhala, 2000) 198–205.

[45] William Grosnick, "The Buddhahood of the Grasses and the Trees," in *An Ecology of Spirit*, edited by Michael Barnes (1990; reprinted, Lanham, Md.: University Press of America, 1994) 197–208.

[46] Howell, *A Feminist Cosmology,* 76.

[47] Warren, *Ecofeminist Philosophy,* 201.

of solidarity; (3) power within, as in drawing on one's inner resources; (4) power against, as in resistance to oppressive forces; and (5) power toward, as in claiming and enacting a positive vision.[48] Buddhist activist Joanna Macy adds a sixth form of power, that of being acted through, as if another larger force is mobilizing one's capacities.[49] A crucial point here is that *any* form of power can be used either destructively or constructively; there are no categorically good types of power. This lack of absolute status requires us to examine carefully *any* use of power to see how, in fact, it is affecting others.

Of the five elements of friendship I have reviewed from Howell, which might be most instructive in working with power? *Liberation from false views* of self is essential for unpacking the Up-Down conditioning, as Warren puts it, that is so draining for women and other oppressed beings. From a nondualistic Buddhist view, it is also critical to unpack how the oppressed person may also abuse power, but in less recognized forms. In other words, no one, whether oppressor or oppressed, is exempt from power abuse. Recognizing this shared commonality, we find that the space opens up for the possibility of becoming friends with one's enemies and thus finding new avenues to peace and the transformation of abusive power structures.

Empowerment or the development of selfhood is the very embodiment of gaining power through mobilizing more resources. In the company of others, one gains confidence, skill, knowledge to strengthen the self as well as the friendship. This cannot be done in isolation, or the sense of personal power may be inflated, not checked by the useful feedback of friendship. Likewise, taking up the *co-created journey* of friendship offers a way to develop skills in "power-with." through acting together on common concerns.

Affection for the "original woman" also relates to the process of empowerment, reinforcing friendships between those who have made similar challenging journeys. But I could imagine the potential for problems arising in judging who is "original" and how original must one be to qualify as an "original woman." Or perhaps competition might arise for the attention of the most original women. Thus, even in the situation of the apparently constructive embrace of power, more destructive forms may surface and be overlooked.

---

[48] Ibid., 199–200.

[49] Joanna Macy, "Faith, Power and Ecology," in *World as Lover, World as Self* (Berkeley: Parallax, 1991) 29–38.

As in friendships with nature, the aspect of *solidarity* holds tremendous potential for shaping friendships that "have the capacity to change the world."[50] Howell cites Mary Hunt in saying that power and politics define women's friendships, calling forth a justice-seeking response. Forming friendships in solidarity, one develops the capacity to confront violence, aggression, and exploitation, meeting one form of destructive social relationship with another based in constructive friendship.

Let me add to these useful elements of friendship two others that may be helpful in relating to power. Karen Warren takes up the long conversation on care vs. justice as a basis for ethics, illuminating some of the historic differences in these classically opposed approaches.[51] She argues for a balanced care/justice ethic, drawing on *both* motivations in engaging friendship and power. This will depend on cultivating the ability to care, including a personal understanding of one's own emotional intelligence, as well as a commitment to justice.

From the Buddhist heritage, I would suggest again the practice of spiritual friendship as providing the opportunity to address power as a moral call. Buddhist organizations such as the Buddhist Peace Fellowship or the International Network of Engaged Buddhists draw on personal spiritual friendships to strengthen their collaborative efforts in opposing structural injustice. Buddhist feminists, anti-war activists, and those working in prison projects incorporate meditation and ceremony into their efforts, adding a practice element to their friendships. This then bolsters a kind of spiritual power that can sustain them in the long commitment to structural transformation.[52]

Conclusion

It has been my privilege to contribute to this conversation on relationality, friendship, nature, and power, building on many fine points developed by Nancy Howell in her feminist cosmology. She has laid out a number of first steps in this dialogue; certainly there is much more to do, as she alludes to with Celie's story: "This is hard work, let me tell you."[53] As she demonstrates, the dialogue between Whiteheadian thought and ecofeminism offers opportunities for commonalties as well as critiques. Likewise,

---

[50] Howell, *A Feminist Cosmology,* 74.

[51] Warren, *Ecofeminist Philosophy,* 107–13.

[52] For examples of spiritual friendship in action, see Christopher Queen, ed., *Engaged Buddhism in the West* (Somerville, Mass.: Wisdom, 2000).

[53] Howell, *A Feminist Cosmology,* 107.

this paper's reflections on Buddhist thought in dialogue with process *and* feminist thought, may shed further light on some of Howell's key concerns.

Perhaps the most significant overlap among all three perspectives is the emphasis on experience as the basis for insight and understanding. There is hardly a shortage on opportunities for experience; thus, invitations to insight only multiply across time. After many hours of thinking about this topic, I conclude that it is important to take the long view, to recognize that the process of liberation and transformation is ongoing. Invited to become friends with beings in nature, we may take some time to figure out how to do that successfully, maybe even a few generations. Invited to work with power through friendship, we may make many mistakes and continue to blame our enemies for our woes.

By taking relationality as the ground and friendship as one possibility within that, we at least have a place to start a conversation. Howell suggests we take up friendship as both "standpoint" and "proposition"—an orientation, a position, a possibility with likely fruitful outcome.[54] This I heartily support—as a feminist, as a Buddhist, and most importantly, as someone living on Earth at this time when so many complex and elegantly evolved relationships are threatened with annihilation. We take up this work as self-corrective practice, learning how to be better partners in relationship with our fellow beings. As we realize how vast this effort may be, humility becomes less a virtue than a necessity. I offer a closing wish of intention: may each small step toward friendship—with nature, with each other, with ourselves—be so deeply rewarding as to encourage the next.

---

[54] Ibid., 88.

6

Does Feminism Need Process?
Yes, No, Maybe, All of the Above

Marjorie Hewitt Suchocki

Nancy Howell's *A Feminist Cosmology*, published in 2000, created a full-length treatment probing the possibilities of dialogue between feminist and process forms of theology.[1] She argued that feminist and process theologians need each other. Process needs the corrective provided by feminism concerning the hierarchical ordering of existence suggested by process, and feminists need the relational metaphysics suggested by process modes of thought.

Howell's work dealt with feminist theology through 1994, the time when Howell actually completed the work for her 2000 publication. But a perusal of writings in feminist theology since that time certainly underscores the relational suppositions woven into feminist writings. Carter Heyward, Sallie MacFague and Rosemary Radford Ruether are the most prominent feminist theologians who to some degree recognize the process suppositions of their work. MacFague specifically refers to her work as "process" in *Super, Natural Christians* and *Life Abundant*,[2] and Ruether uses process thought in *Gaia and God*.[3] Anna Primavesi in her 2000 *Sacred*

---

[1] Nancy R. Howell, *A Feminist Cosmology: Ecology, Solidarity, and Metaphysics* (Amherst, N.Y.: Humanity, 2000).

[2] Sallie McFague, *Super, Natural Christians: How We Should Love Nature* (Minneapolis: Fortress, 1997); and idem, *Life Abundant: Rethinking Theology and Economy for a Planet in Peril* (Minneapolis: Fortress, 2001).

[3] Rosemary Radford Ruether, *Gaia and God: An Ecofeminist Theology of Earth Healing* (San Francisco: HarperSanFrancisco, 1992).

Gaia also refers to process thinkers as compatible with feminism,[4] as does Grace Jantzen (albeit in less sanguine fashion) in her 1999 *Becoming Divine.*[5] Thus, there seems to be ample confirmation for, at least, the feminism-needs-process thesis that Howell develops.

Despite such confirmations that apparently answer "yes" to this question, in this essay I will investigate the "No," "Maybe," and "All of the Above" options listed in my title. And of course the opposite inquiry— does process need feminism?—will likewise be explored in this ambiguous manner.

NO! or, Feminism Stands Alone

Since feminism is, after all, a very relational way of speaking about the human condition, it may seem a bit contradictory to suggest that feminism stands alone. But historically, feminism has struggled to extricate women and women's issues from their submergence in patriarchy. I am reminded of those Michaelangelo sculptures in Florence leading up to that magnificent homage to maleness, Michaelangelo's David. Each of the preceding sculptures appears to be unfinished. That is, we see in each presentation the process as well as the product. A figure is shown as it partially emerges from the stone, as if it were embedded within, and is now—before our very eyes—struggling to become free through the medium of the sculptor's knife. We imagine the rest of the figure hidden inside the as yet uncarved stone and know that the figure has been trapped in the stone for ages, just waiting for Michaelangelo's liberating artistry. Women, likewise, have been buried within the seemingly solid stone of patriarchy. In our immediately past century, the liberating knife was taken up by Simone de Beauvoir in 1949,[6] followed by Betty Friedan in 1959,[7] and then with increasing fervor by the feminists of the 1960s—Mary Daly and Rosemary Radford Ruether being the most prominent religious thinkers among them.

Of course, no movement springs without antecedents in human history. One could point to male sculpting tools forged by persons such as John Stuart Mill—and of course there were the foremothers throughout

---

[4] Anna Primavesi, *Sacred Gaia: Holistic Theology and Earth System Science* (London: Routledge, 2000).

[5] Grace Jantzen, *Becoming Divine: Toward a Feminist Philosophy of Religion* (Bloomington: Indiana University Press, 1999).

[6] Simone de Beauvoir, *Deuxième sexe* (Paris: Gallimard, 1949), translated as: *The Second Sex,* trans. H. M. Parshley (New York: Knopf, 1953).

[7] Betty Friedan, *The Feminine Mystique* (New York: Norton, 1963).

history, who also wielded their own pressure upon the stone of patriarchy, particularly those of the nineteenth and early twentieth century—
Charlotte Perkins Gilman, Jane Addams, Virginia Woolf! Those who
worked in our own time built upon the work of women and some men
who went before them. But today, were it not for the work of Simone de
Beauvoir, Betty Friedan, Mary Daly, Rosemary Ruether, and the growing
legions of women who joined them in the work, would we women not be
buried still within the stone?

The impetus for their efforts came from their passionate commitment to the full dignity and equality of women. This, and this alone, fu-
eled their labors. To suggest that they "needed" a metaphysic to support
their work would seem to fly in the face of their non-metaphysically based
heroism. These foremothers of ours fully recognized the interdependence
of persons and systems—one has only to read their analysis of the social
snaring created by patriarchy to realize they were fully aware of the power
of relations. But knowledge of this power did not require an analysis of the
metaphysical basis of this power in order to allow them to do their work.
Rather, they simply used relation—working together, both cooperatively
and critically—to accomplish the common task.

It was this very working together that produced the slow recognition
that oppression is a hydra-headed thing and that patriarchy is reinforced
by racism, classism, nationalism, heterosexism, and any other "ism" that
denies the full personhood of another. Social location emerged as a critical
element in feminist work, along with the self-and other-critique concerning racist oppressions, which worked to the benefit of whites in general,
including white feminist professionals. Revisionary works and new explorations from various social locations such as womanist and Mujerista and
bamboo theologies began to appear. White feminists wrote "mea culpa"
monographs, and feminism as a whole began to become self-critical.

Did feminism need process metaphysics in order to facilitate this history? Of course not! The work proceeded through deeper and more probing analyses not only of sexist dynamics, but of the constructive implications for developing feminist alternatives. To suggest, then, that feminist
theology needs process theology with its metaphysical explications to further the work of feminism would seem, on the face of it, to be little more
than a belittlement of feminist work. Women's work stands on its own; it
does not "need" the reinforcement to be derived from a system that, after
all, has been developed most fully by far too many white professional men!
Women process thinkers have been few and far between! Is it not an out-
rage to suggest that feminism now needs process thinking to supplement

the hard-won work of feminists alone? Thus, one can imagine a resounding NO to the question, Does feminism need process?

But Does Process Need Feminism?
NO, or, Feminism Adds Nothing to Process!

Whitehead wrote *Process and Reality* for the Gifford Lectures of 1929.[8] And while there is certainly evidence that Whitehead actively supported the women's rights movement of his time, their issues dealt largely with women's right to vote. Appreciation for Jane Addams at Hull House certainly would have been in keeping with Whitehead's stance, but contemporary feminism was something beyond Whitehead's own time. It would be anachronistic to suppose that he should have or could have incorporated feminist insights into his development of *Process and Reality*.

And indeed, that work is not particularly concerned with social issues. Instead, Whitehead seems to ask the question, "What must existence be like at its core, given the internal relations and responsive nature of even tiny sub-atomic elements discovered by physicists?" He wanted to re-think the nature of our whole universe in terms of the centrality of responsive internal relationships.

And so, of course, he did, using all the data he could muster, beginning with personal experience. Social history, religion, physics, mathematics, and the history of philosophy all fed into his analysis. But he felt hampered by the non-dynamic implications of most philosophical language and even scientific language. Up to his time, most philosophies and Newtonian physics had been oriented around notions of substance and accidents, with a particular primacy given to the data of sensory experience. But Whitehead needed to go beyond that if he was going to describe a world that was deeper than substance and accidents and even deeper than sensory data. He needed a language that moved, reflecting the restless advance created by innumerable processes of entities that come into being by feeling the past, integrating that past into a new present, and demanding an accounting in the future that ever succeeds each one. Existence is not static, but dynamic; not substance, but process; not limited to what we can know through our sense organs, but deeper still in the connections between our embodied selves and the wider world. So, of course, he made up his own language, which continues to drive persons either insane or away

---

[8] Alfred North Whitehead, *Process and Reality*, 1929, corrected ed., ed. David Ray Griffin and Donald W. Sherburne (New York: Free Press, 1978).

from his insights, back to the safer world of sense data and/or analysis of how we use words, in endless rounds of repetition of the obvious.

So Whitehead forged a basic model. It begins, naturally, in the middle of his story, for he is describing the world as we experience it, and we never begin from nothing. We always begin from a standpoint that presumes a past. That past, he said, pulsates with throbs of energy pounding upon the door of its future in every millisecond. So we begin, or emerge, through the force of an insistent past that demands our becoming. We begin by feeling the force of a multitude of entities composing our influential past. These entities are a bit cacophonous, and it is our task to become ourselves through dealing with the cacophony of the past, unifying it into our own new becoming. How we are able to do this drove Whitehead, willy nilly, to a notion of God; for apart from some guiding aim as to what one might do with the past, there would be sheer chaos. In order to account for order, there must be a highly complex multi-directional force that could be felt by every entity whatsoever at the very beginning of its emergence.

So Whitehead figured that there are two creative sources necessary to account for any and every existence: a past that pushes a new present into becoming, and a future that calls a new present into its own particularity. In a sense, the present is created through the pressure of two "bookends," the past and the future. But remember, we are talking about a dynamic model of internal relations. The present becomes itself as it internalizes its past and its future. Plato once coined a remarkable phrase that Whitehead adapted to describe such a situation: the present is that which is always becoming, and never really is. Once its dynamic process of self-creation is over, it joins the past, calling for its own successors. We are always moving on.

So, then, here in several paragraphs I have inadequately summarized 353 pages of a complicated text. Clearly there is much to study to fill in the blanks of my cursory description, but hopefully this is enough for you to get a sense of Whitehead's dynamic accounting of existence. He called his work a cosmology, or a description of the processes of becoming that apply to any and every element in the entire universe.

It is one thing to sit in one's armchair and think such great thoughts, but Whitehead was in some sense a pragmatist. He saw his model as one to be tested, adjusted, adapted, and changed according to whatever new data from whatever source showed up on the horizon. In a profound sense, these new data are peculiarly to be supplied from the study of physics. If, for example, physicists should discover that those seeming elements of ceaseless change described by Whitehead were not the most basic reality, that underlying them was a still more essential thing that was far more akin to substance than to dynamism, then his system would have to be seen as provisional, and a new cosmology would have to be developed.

Because Whitehead's system addresses a cosmological structure of existence that underlies all forms of organic and inorganic systems, there is a real sense in which changes in social systems are irrelevant to his analysis. Empires may grow and fall, liberations succeed or fail, social systems thrive or decay, and none of these monumental facts of historical existence challenge, correct, or confirm the fundamental cosmology. In this sense, then, the question of whether Whitehead needs feminism is a question that simply misses what his system does. It's something like asking whether a rectangle needs a particular system of government. Rectangles are qualitatively different things from governments, and the question simply makes no sense. Likewise, the Whiteheadian cosmology simply describes the dynamics of existence per se; it doesn't "need" any particular form of social existence—to the contrary, it underlies all of them. Even though it has deep implications for how we should organize and structure our personal and social realities, the basic model is what it is regardless of whether we do or do not follow its implications.

So then, at a basic level, we can look at both feminism and Whiteheadian metaphysics, and state categorically that each can do very well on its own, thank you. Neither particularly needs the other. But now let's move into the "maybe" of my title, and explore these matters a bit further.

On the Other Hand, Maybe
Process Thought Could Benefit by Feminism

While the things I have explained are correct, there is, however, another aspect of Whitehead's thought that changes things a bit. It is so that only contradictory data from physics could deeply undermine Whitehead's cosmology. And lest you draw in your breath with too much hope that maybe this great system could indeed be undone by a century of progress in the study of physics, I must tell you that currently there are outstanding physicists who are quite interested in Whitehead's analysis. So far, it seems to fit as a model of the way the world is put together at its most fundamental level.

But Whitehead was not content simply to draw up a cosmic map of how things are. His own metaphor for the way one works with metaphysics was that of an airplane—or "aeroplane" in the talk of the 1920s. One develops a scheme, based on the best evidence at hand, but a cosmological scheme all by itself is something like an airplane that never takes off. The plane is as good as its usefulness. Even so, a cosmological scheme is as good as its usefulness. One must "fly" it to various territories, test out its implications and applications, and tinker with it until it straightens up and flies right, so to speak.

Whitehead himself certainly applied his theories to fields beyond his native mathematics and physics. Most notably, he applied his theories to education, to religion, and to the area of social history by seeing if the dynamics of process could give some accounting of the adventures of ideas throughout human history. And it is also certainly the case that the folks at the Center for Process Studies have busily flown that Whiteheadian airplane into areas as diverse as biology, theology, economic theory, politics, psychology, and—like Whitehead—education.

This is to say that a cosmology is as useful as its application. To every cosmological scheme, one should ask, "so what? What difference does this accounting of experience make to the way I live my life, shape my values, formulate goals, create societies?" And since a cosmology is by definition to be universally applicable, there is in principle no place where the airplane should not be able to land. The cargo on this plane is the package of insights into the thoroughly relational nature of all existence. It might be the case, then, that the process plane needs to land into feminist territory in order to test its aerodynamic worthiness as well as the value of its cargo in that proudly independent but nonetheless relational world of feminism.

Howell suggests that if process thinking were tested against the insights of feminism, process thinkers would discover and correct a bent toward hierarchical value schemes built into the process system. Such hierarchical tendencies, she is quick to admit, are not essential to the fundamental Whiteheadian analysis of relational existence. But persons working with Whitehead's system have injected their own predilection for hierarchy into their use of process thought. Landing the plane smack dab in the midst of feminism would jolt such persons into awareness; they would see the error of their ways, and correct the course for their subsequent flights.

But I am not so sure this is what process gains from feminism. Even ecofeminists, of course, tend to slap mosquitoes, and are glad to take an antibiotic or two to keep pesky invaders at bay. We value diversity, but we'd be pleased to eliminate the particular diversity represented by organisms responsible for AIDS, smallpox, tuberculosis, cholera, and a host of other devastations. And if a ravenous wolf were about to leap on a small child, we'd do all in our power to save the child, even if it meant destroying the wolf. When push comes to shove, we tend to experience what David Ray Griffin would call a "hard core common sense" assent to hierarchies such as these.

In a process world, the justification for such hierarchies is "intensity of experience," which itself follows from self-enjoyment at increasing levels of complexity. Each and every unit of experience, whether human or otherwise, is intrinsically valuable in and for itself; there is no unit of experience whatsoever without some degree of intrinsic value. The degree is based on the ability of the unit to incorporate diverse influences from its past into the unity of its own becoming. Because we humans live in and through our very complex bodily systems, we are capable of incorporating more contrasting influences within us than is the case for a mosquito. Hence human life presumably entails greater enjoyment, intensity, and complexity than is the case for nonhuman life. However, the difference is one of degree; there is no sharp distinction between forms of life, since all are interdependent and interconnected.

In addition to the intrinsic value of every unit of experience, there is also instrumental value, or the value that one contributes to the newly becoming universe. Bluntly, the food you last ate had intrinsic value in its own right—including enjoyment of experience in a Whiteheadian world—and instrumental value for you, contributing to your own intrinsic value. And your value, in turn, is not a terminus point as if you—or humanity per se—were the height of all intrinsic value; you yourself—and humanity per se—are of instrumental value insofar as we contribute our own enjoyment to the enjoyment of others. In short, the value system of a Whiteheadian world is a giving and a receiving. The hierarchy of values is not absolute, but relative to the greater whole, and relative to the degree of complexity attainable by each individual unit of experience. Each unit exists in responsible relation to all others.

Given this system, biologist Charles Birch and theologian John B. Cobb Jr. wrote a book called *The Liberation of Life* in which they argued for ecological responsibility.[9] As Nancy Howell points out, their work "balances respect for each entity's intrinsic value with its instrumental value for others."[10] But she cautions us against their development of gradations of

---

[9] Charles Birch and John B. Cobb Jr., *The Liberation of Life: From the Cell to the Community* (Cambridge: Cambridge University Press, 1981).

[10] Howell, *A Feminist Cosmology*, 55.

value that judge some things as less intrinsically valuable than others. For instance, despite the intrinsic value of all existence, the greater capacity of humans for richness of experience suggests that humans are of greater value than mosquitoes. The higher up the complexity chain a being is, the greater value it has over those below it. Boundaries between species are indeed blurred, and interdependence through instrumental value certainly obtains, but there is a clear basis for determining importance of existence.

It's not the case that Birch and Cobb ignore the importance of instrumental value in determining placement on this chain. Plankton, for example, presumably has little intrinsic value in its particular form of existence, but its instrumental value for other life forms more than compensates for its low intrinsic value. The relational hierarchy, then, is a complex thing involving intrinsic and instrumental value in a dynamic flow where there can be no "finality" placed on any particular construct of value.

For all this, Howell claims that this process "liberation of life" deeply needs a feminist corrective. The hierarchy, however blurred, that deems complexity and richness of experience to be a fundamental criterion of value would seem to be all too useful to those who would claim that men are far more complex and rich in experience than any other life form, particularly women. Haven't women experienced societies ordered on just such principles for far too long? Thus, Howell argues that process needs feminism in order to correct its dangerous lurching toward that great chain of being that culminates in the almighty male.

Howell's correction attempts to keep a balance between the individual and communal aspects of being/becoming. Intensity, she notes, comes about through the capacity to hold diverse elements together within the unity of experience. Therefore, the greater the diversity, the greater the intensity that is possible. This importance of diversity guards against the tendency to make intrinsic experience alone the determiner of value. She buttresses her argument by noting that within process theism, God receives every unit of existence into the divine nature, there to integrate it within the everlasting divine becoming. The diversity of finite experience therefore enriches the divine experience through its continuous introduction of complexity, and therefore intensity. Hence, without diversity, intensity is diminished, and richness of experience is also diminished. Hierarchies, then, are undermined. Societies built on such a model would take particular care to encourage diversity within the community, thereby increasing intensity of experience not simply for some, but for all.

I have no quarrel with Howell's wish to undermine absolute hierarchies of existence. However, her appeal to the necessity of diversity to divine experience seems to reposition the very ladder she seeks to destabilize. Isn't the appeal to divinity the appeal to the top of the ladder, so to speak? And while the impetus for Howell's correction of Birch and Cobb comes from Howell's feminist experience, I note that she develops the corrective from within the parameters of process metaphysics. While she appreciatively delineates various ecofeminist positions against hierarchy, she does not use their arguments as she reconstructs process in a manner more agreeable to ecofeminist sensitivities.

This, then, would seem to buttress my earlier position that process does not actually need feminism, whether eco or any other, in order to go its merry way. While I think Howell is on the right track, I suggest that her biologically oriented analysis be supplemented with a more socially oriented analysis. That is, Birch and Cobb, as well as Howell, note the shadings and complexities of value in the biological realm. The necessary combination of intrinsic and instrumental values along with complexity and intensity of experience means that all hierarchies are contextual and multiple. There can be no single hierarchical ordering, but only combinations of combinations of such ordering. My earlier illustration of the value of plankton makes this obvious: a simple hierarchy of intensity of experience puts plankton very low on the list—but a complex hierarchy that looks at the sustaining environment necessary for intensity of experience puts plankton high on the list. Complex relational judgments are required in the biological world.

But the same is true in the social world. Problems emerge not from hierarchical structures per se, but from simplistic, one-dimensional structures that fail to observe the multifaceted and ambiguous edges attendant upon a relational ordering of society. In 1911 Charlotte Perkins Gilman wrote a utopian novel, *Moving the Mountain*, in which she pays due homage to the multiple hierarchies that are necessary for the creation of society.[11] Farmers, garbage collectors, and cooks are at the top of the hierarchy in some respects, while in other respects artists play a pivotal role. Hierarchies shift and change; there is no single absolute hierarchy that appropriately determines the value of all others. Men and women in a society based on pluralistic hierarchies are not judged according to gender, but as persons-in-community, each of whom contributes values according to a variety of scales. The problem is not with hierarchy per se (we continue to

---

[11] Charlotte Perkins Gilman, *Moving the Mountain* (New York: Charlton, 1911).

take antibiotics and to slap those mosquitoes) but with a single hierarchy with the pretense of being absolute. Pluralistic hierarchies recognize the multiple roles played by all members of the society, and hence the relative nature of all hierarchies. A single hierarchy absolutizes its roles; pluralistic hierarchies relativize all roles.

So how does process benefit from feminism? While the resources for acknowledging the necessary multiplicity of hierarchies exist within process itself, the impetus for discovering this comes from the feminist experience of climbing out from under the oppressive single-hierarchy value system of patriarchy. "Consciousness raising" was an early feminist strategy of the 1970s, waking women up to look at their role in society. The result of this awakening was an impetus to begin chipping away at that marble within which women were trapped. Even so, feminism has provided a "consciousness raising" function for process thinkers as well, persuading us to look at the way patriarchal assumptions can twist even the radical relationality of process into un-processive social hierarchies.

And Maybe Feminism Benefits from Process?

In a sense, this question is already answered positively through the work of those feminist thinkers I have already cited who pull process ways of thinking into their feminist theologies. Single-hierarchical worlds are based on absolutes that themselves derive from substance metaphysics. And the ideal of substance as "that which requires nothing other than itself in order to exist" pervades patriarchal thinking. Long ago Nietzsche pointed out that it was difficult to use the master's tools to dismantle the master's house. Like patriarchy, non-relational assumptions have so dominated our ways of thinking that we hardly notice them. But non-relational thinking has been the metaphysical foundation of patriarchy. If feminists are to dislodge patriarchy, they must also dislodge its foundations in non-relational thinking.

To some degree this happens in and through the feminist attack on dualistic thinking. Feminists point out the oppositions that mark patriarchal thinking, such as subject/object, mind/body, man/woman, God/human, human/natural. However, feminists have not necessarily recognized that these dualisms are themselves dependent upon a worldview based on the primacy of self-contained atomic substances. The binary oppositions of substantive thinking do not admit of ambiguity, or the sense in which each polar opposite depends upon its other. A "substance" worldview is an either/or worldview. Thus feminists must address themselves not only to the polar oppositions, but also to the substance metaphysics that supports these oppositions. It is as if patriarchy stands upon a substantive rug; if we attempt to remove patriarchy without removing the rug, we invite some other form of dualistically prone system to stand in its place. We must pull the rug out from under patriarchy, and a process metaphysics of relationality could be very helpful in this endeavor.

Feminists can talk about relationality as the basis of existence without recourse to process thinking—this is evident as early as in Carter Heyward's 1982 *The Redemption of God*.[12] She presupposes a fundamentally relational world. However, the substance way of thinking is so engrained in us that simply asserting relationality is not always a sufficient way to dislodge these long habits of thought. One must go beyond assertions to arguments based upon an analysis of as broad a range of experience as possible. This is precisely what process does.

In a sense, the century-old philosophical trend of turning from an analysis of experience to an analysis of language undercuts the attempt to undermine the metaphysics of patriarchy. The danger in this is that because metaphysics is not a "fashionable" way of philosophic thinking, it is not an examined way of philosophic thinking. Thus substance assumptions quietly continue to force their unexamined power on our personal and social lives. So long as feminists are kept away from examining the foundations of patriarchy in substantive thinking, they will find themselves fighting the hydra-headed monster. No sooner is one head lopped off than another takes its place. Substantive thinking generates dualistic thinking, and dualistic thinking is the static, non-relational, single hierarchy of patriarchy. We must get rid of the monster, and not just the heads. A relational worldview that totally displaces the static worldview can be most helpful in this task.

So, Then: All of the Above?

And so I come to my ambiguous answer to the questions posed at the beginning of this essay. Does feminism need process thought? Does process need feminist thought? Yes, no, maybe, or all of the above? My ambiguous answer is that clearly, we should opt for all of the above! On the one hand, there is the "yes": Both feminism and process are relational ways of thinking and, by definition, value interdependence. We not only value interdependence, we think it is the fundamental way of things: we exist in

---

[12] Carter Heyward, *The Redemption of God: A Theology of Mutual Relation* (Washington, D.C.: University Press of America, 1982).

an interrelational world. If so, then process and feminism, while discrete systems, are interdependent, and therefore need one another. To deny this need is to assert a self-sufficiency that belies our relational assumptions.

And yet historically, there is also the "no." Each system actually developed in quite different social locations. Process thinking arose as an alternative philosophical construction of the world; feminism arose as an alternative social construction of the world. Each had its own methodology; each generated its own school of thoughtful activists and active thinkers. To use the process terminology introduced earlier, each has intrinsic value regardless of the other. But each also has instrumental value that affects the other, which led to my "maybe." It may well be that more explicit attention to the other on the part of feminists and process thinkers will strengthen the work of both. We are both, after all, in some respects on the margins of academic and social acceptability. While we might well take pride in the idiosyncratic status this gives us, we might also increase our instrumental influence if we learn from one another and, to some extent, join forces, while each remaining ourselves. And who knows? Perhaps such an alliance will—to combine my metaphors!—not only chip away more of that marble, it might also yank that patriarchal rug right out from under society's feet! So, then, does feminism need process? Does process need feminism? Yes, no, maybe, and all of the above!

7

Beyond *A Feminist Cosmology*

Nancy R. Howell

EVEN (or especially) as the author of *A Feminist Cosmology: Ecology, Solidarity, and Metaphysics*, I plan to engage the book critically in the following essay. *A Feminist Cosmology* is a product of its time, which means that my theological reflection enjoyed the benefits of important feminist and ecofeminist models that were groundbreaking in the 1980s and 1990s. While I am honored to identify with and write from within the maturing feminist theological and Whiteheadian traditions, I understand that my ideas were circumscribed by the limits of my own theological imagination. In repaying the debt that I owe to senior scholars who inspired my thinking and to the authors who have engaged *A Feminist Cosmology*, I compose my current reflections as an imaginative, heuristic proposal. If others find the following possibilities interesting, I hope that my suggestions reciprocate the continuing inspirations that move my scholarship forward, and otherwise I hope the proposal keeps me accountable for greater depth, creativity, and clarification of ideas central to justice for women, nature, and others at the margins.

Proposal 1:
Attention Epistemology and Narrative

The theme that guides my proposal is a quest for the growing edges in ecofeminist theology, which appropriately and appreciatively follows review of the valuable deconstructive and constructive work of feminist eco-theologians. We honor the theological writing of groundbreakers, such as Rosemary Radford Ruether and Sallie McFague, when we incorporate or embody their contributions in our work, but support their transformative project with new questions and insights.

Standing on the achievement of Sallie McFague, I draw particular attention to the attention epistemology, which is carefully described in McFague's *The Body of God: An Ecological Theology*.[1] Attention epistemology requires refocusing so that the main attention of scholarship and observation moves from self-interest and other-utility to another creature, one who possesses particularity and value apart from the perspective of and usefulness to the observer. A methodological approach that entails attention epistemology requires more than selflessness on the part of the observer and, in fact, demands heightened self-awareness in order that the observer may sort out the perspective, presuppositions, and biases that he or she brings to the observation of the other creature, who may be radically or modestly different from the scholar. If the observer can achieve the standpoint of attention epistemology, then the scholar may be truly observant in allowing the data, anecdotes, and observations to speak for themselves. Perhaps attention epistemology is the method by which scholarship may approach the strong objectivity advocated by Sandra Harding, because attention epistemology demands self-critical acknowledgment of ones standpoint (which Harding's standpoint epistemology describes). Turning one's gaze toward the observed, after clearly differentiating self from other, means that the scholar may engage in "listening to another, the other, in itself, for itself."[2] McFague connects attention epistemology to embodiment and action, because embodiment is the standpoint from which we know ourselves as differentiated and others as different from us:

> An attention epistemology is central to embodied knowing and doing, for it takes with utmost seriousness the differences that separate all beings: the individual, unique site from which each is in itself and for itself. Embodiment means paying attention to differences, and we can learn this lesson best perhaps when we gauge our response to a being very unlike ourselves, not only to another human being (who may be different in skin color or sex or economic status), but to a being who is *in*different to us and whose existence we cannot absorb into our own—such as a kestrel (or turtle or tree). If we were to give such a being our attention, we would most probably act differently than we presently do toward it—for from this kind of knowing—attention to the other in its

[1] Sallie McFague, *The Body of God: An Ecological Theology* (Minneapolis: Fortress, 1993).
[2] Ibid., 49.

own, other, different embodiment—follows a doing appropriate to what and who that being is.[3]

Behind McFague's description of attention epistemology is the assumption that knowledge affects action. If McFague's assumption is correct (and I hope it is), then attention epistemology is a way to assure that particular knowledge is more reliable, because the scholar's standpoint is less self-absorbed and more attentive to the particular other. Embodied knowing not only appreciates the intrinsic value of the other, but is moved to action by the other. The knower is transformed into one who loves the other and acts on behalf of the other.[4]

While McFague refers to Barbara McClintock's "feeling for the organism" as an example of attention epistemology, my research takes me to Jane Goodall and her work with chimpanzees. Arguably, Goodall's approach to chimpanzee observation is a kind of attention epistemology, a risky methodology within the sciences, as Goodall defiantly admits:

> As I got to know [chimpanzees] as individuals I named them. I had no idea that this, according to the ethological discipline of the early 1960s, was inappropriate—I should have given them more objective numbers. I also described their vivid personalities—another sin: only humans had personalities. It was an even worse crime to attribute humanlike emotions to the chimpanzees. And in those days it was held (at least by many scientists, philosophers, and theologians) that only humans had minds, only humans were capable of rational thought.[5]

Similarly, Goodall recalls struggles to publish within the scientific community.

> The editorial comments on the first paper I wrote for publication demanded that every *he* or *she* be replaced with *it*, and every *who* be replaced with *which*. Incensed, I, in my turn, crossed out the *its* and *whichs* and scrawled back the original pronouns. As I had no desire to carve a niche for myself in the world of science, but simply wanted to go on living among and learning about chimpanzees, the possible reaction of the editor of the learned journal did not trouble me. In fact I won that round: the paper when finally published did confer upon the chimpanzees the dignity of their

---

[3] Ibid., 50–51.

[4] Ibid., 50.

[5] Jane Goodall and Philip Berman, *Reason for Hope: A Spiritual Journey* (New York: Warner, 1999) 74.

appropriate genders and properly upgraded them from the status of mere 'things' to essential Being-ness.[6]

Goodall violated conventions of ethology by naming animals (instead of numbering subjects), by referring to animals with personal (*he, she*) and relative (*who*) pronouns usually reserved for human persons, by describing the personalities of animals, by asserting that chimpanzees are rational animals (endowed with minds), by interpreting animal behavior in terms of emotions, by interpreting motives or purposes in animal behavior, and by using vocabulary (*childhood, adolescence*) designated for human life cycles to describe chimpanzee life stages.[7] While Goodall's observation and reporting method broke scientific conventions, note that her approach assumed the intrinsic value of chimpanzees and engaged the particularity of the species and of individuals, whose family relationships and histories fill her publications and films. Goodall's accounts of her own life indicate that the Gombe experience with chimpanzees transformed her life, and *Reason for Hope* is a memoir about Goodall's spiritual journey among the chimpanzees. The intrinsic value of chimpanzees has also encouraged Goodall to activism on their behalf. The evidence of Goodall's activism includes the founding of The Jane Goodall Institute for People, Animals and the Environment, which supports "research and conservation projects involving chimpanzees and other wildlife in Africa, to improve the conditions of chimpanzees and other animals in captivity, and to raise awareness and understanding of these issues."[8] Goodall and Marc Bekoff together founded Ethologists for the Ethical Treatment of Animals/Citizens for Responsible Animal Behavior Studies to protect animals from cruel or unnecessary treatment as subjects of research projects.[9] I draw the conclusion that Goodall and her research are shaped by attention epistemology, which recognizes the intrinsic value and particularity of chimpanzees and which means that her own being and activism are intimately tied to her love for chimpanzees.

Perhaps, attention epistemology (as exemplified in Goodall's observation of chimpanzees) makes a connection with narrative theology. Attention epistemology is a method by which humans may hear the stories

---

[6] Goodall, *Through a Window: My Thirty Years with the Chimpanzees of Gombe* (Boston: Houghton Mifflin, 1990) 15.

[7] Ibid., 14.

[8] Ibid., 281.

[9] Goodall, Forward to *Minding Animals: Awareness, Emotions, and Heart,* by Marc Bekoff (Oxford: Oxford University Press, 2002) xiv.

of nonhuman animals and nature, in the way that Goodall "listens" using an observer's eye. Attention epistemology creates a space for animals to communicate using voice, movement, and relationships (and I intentionally associate communication with more than spoken stories in order to include McFague's concept of embodied knowing and Whitehead's concept of feeling/prehension among the ways we know each other and ourselves). The relationship of observer and chimpanzee enters human experience, becomes part of the human story, and creates new insight—compelling insight that calls for a more adequate ontology (such as Alfred North Whitehead's relational ontology or Mary Daly's feminist ontology) and a corresponding liberative praxis. Because attention epistemology accesses the potential to know chimpanzees as a community and as individuals, the chimpanzee "narrative" then suggests the promise that community may be realized between humans and nonhuman animals, a lesson that humans should extend to just and liberating relationships with each other. Attention epistemology enlarges the human narrative by engaging the stories of specific, diverse humans and animals.

In sorting through the methodological value of attention epistemology and narrative, I am especially interested in how new avenues of ecofeminist thought might be opened. Much ecological theology in general is drawn to questions about and discussions of the ecosystem, and some ecological theology develops arguments using scientific evidence from particular species. What has not happened in ecological theology or ecofeminist theology is a systematic examination of a single species, including attention to diversity of individuals within the species.[10] *A Feminist Cosmology* is yet another example of ecofeminism concerned about the big picture in nature (the whole) rather than the creatures (the parts) whose species stories might teach us something about nature and animals. Seeing the tendency in my own work to focus on the whole and to neglect the parts, I noticed that I often cited anecdotes or data about particular species to support points arising from generalizations in science and philosophy, and I had not really challenged myself to face the body of empirical evidence about one species that might ultimately challenge the neat Whiteheadian and feminist formulations characteristic of my thinking. To balance my emphasis on "the whole" ecosystem in *A Feminist Cosmology,* I am current-

---

[10]   One noteworthy exception to my generalization may be the work of Carol Adams, who wrote very systematically about the relationship of meat-eating and gender bias. Her book is very specific about the lives of animals exploited in the mass production of meat and poultry; Carol J. Adams, *The Sexual Politics of Meat: A Feminist-Vegetarian Critical Theory* (New York: Continuum, 1990).

ly undertaking a detailed empirical and theological reflection on chimpanzees—one "part" of nature (with the awareness that I might have selected
any species for the task), which raises for me questions about the nature
of the soul, sin, culture, and nature, as well as theological concepts of God
and animals that might be adequate to the empirical, ethological revela-
tion of the lives and stories of chimpanzees.

*My first proposal for the future of feminist and ecofeminist theology is
that we explore the promise of attention epistemology and narrative as methodological approaches to both whole and parts in nature, so that we might
balance our generalizations bout the ecosystem with insight from the depths
of species and individual lives of animals or other creatures in nature.* Such
an approach reflects my growing interest in science-religion dialogue, in
which science provides a window on the intimate and particular stories of
chimpanzees. By engaging the scientist's stories (data and observations) of
chimpanzees, I anticipate that my ecofeminist theology will move more
dramatically from romantic understanding of nature and abstract understanding of humans to confrontation with both the delight and tragedy in
human and nonhuman nature.

Proposal 2:
Coalitions and Complementarity

A Feminist Cosmology attempts to demonstrate that Whiteheadian and
feminist claims are complementary. At the time of writing, I thought I was
developing a very simple argument for the rationality and appropriateness
of developing a relational feminism using Whiteheadian philosophy and
theology. In so doing, I was naming what could be called a "Claremont
School of Feminism" already begun in the work of Marjorie H. Suchocki,
Catherine Keller, Rita Nakashima Brock, and a number of other feminists.[11] While I suggested that Whiteheadian philosophy might be a helpful schema for systematizing some feminist thought, I was (and still am)
unwilling to argue that feminism must always be Whiteheadian. Likewise
not all Whiteheadian thought is likely to be feminist, although I would
like to encourage all who write Whiteheadian theology or philosophy to
seek justice for women.

---

[11] At a conference in Claremont in May 2004, the term *Claremont School of Feminism*
emerged in a conversation with Rita Nakashima Brock as we discussed the unique relational approach of Whiteheadian feminists who studied at Claremont Graduate School
(often with John B. Cobb Jr. as dissertation advisor).

While some scholars integrate Whiteheadian and feminist thought, the activist dimension of my scholarship seeks coalitions with other scholars who advocate justice for women and nonhuman animals and nature. One goal of *A Feminist Cosmology* was to write in serious dialogue with scholarship by women of color. While I am fully aware that white feminist dialogue with scholarship by women of color is frequently very difficult and that my particular attempts at dialogue may be awkward and even painful, I made an intentional commitment to engage womanist, Asian feminist, and *mujerista* theology for two reasons. The first reason was that I did not want to write as if I have not been deeply affected, challenged, and transformed by the scholarship of women of color. My intention was to reciprocate and acknowledge the ways in which my ideas have been shaped. The second reason that *A Feminist Cosmology* engaged scholarship by women of color is that the coalitions that I hope to build for the sake of justice toward women require that my ideas be tested in terms of their adequacy for diverse women. I resisted using diverse women's scholarship to support my preconceived scholarship and instead shaped and evaluated my ideas in response to Chun Hyun Kyung, Patricia Hill Collins, Kelly Brown Douglas, bell hooks, Ada María Isasi-Díaz, Kwok Pui-lan, and Delores Williams (among others who were not cited specifically in the book). As a result of publications by these women scholars, I gained a new perspective and will never see the world or interpret events in the ways that I did before experiencing their ideas.

Coalition-seeking promises to generate new insight and perspective in the case of multireligious and multicultural dialogue, as well. Buddhist-Christian dialogue, for example, has influenced my reflection about eco-feminism and chimpanzees. Japanese culture and religion support a different perspective on observation of chimpanzees. In 1985, Jun'ichiro Itani noted that "Japanese culture does not emphasize the difference between people and animals and so is relatively free from the spell of anti-anthropomorphism . . . we feel this has led to many important discoveries."[12] Not encumbered by the sharp division that Western culture developed and maintained in a Great Chain of Being that separated humans and animals, Japanese culture depicted in folk tales and poetry the monkey as mirror of humanity.[13] Japanese scholars considered animal primates to be as individually diverse as humans. Much like Goodall's approach, the Japanese

---

[12] Cited (without exact attribution) in Frans de Waal, *The Ape and the Sushi Master: Cultural Reflections of a Primatologist* (New York: Basic, 2001) 85.

[13] Ibid., 190.

approach to primatology included naming animals, distinguishing them by physical markers, noting distinct personalities, and documenting kinship and relationships, including friendships, rivalries, and rank.[14] In the 1960s, Itani and Toshisada Nishida studied chimpanzees in Tanzania, and, in keeping with the approach of Japanese primatology, Itani and Nishida observed the chimpanzees, presupposing that the animals are highly social beings and that survival depends on social connection with the group. The results of the chimpanzee study in Tanzania established that chimpanzee groups are highly stable male-philopatric social communities, with migration of females rather than males between groups of chimpanzees.[15] Note that Jane Goodall's approach to chimpanzee observation shares some practices with the Japanese approach, and, initially, both Goodall and the Japanese scientists were criticized by Western scientists. Eventually, however, Western science realized that the Japanese perspective was significant in providing important bodies of data and crucial theories.

My second proposal for ecofeminist and feminist reflection is that we seek coalitions that give us insight and perspective and that support the justice that we advocate. I continue to agree with Janice Raymond that *worldly separatism* is a viable option, which permits women (and others by analogy) to come to voice and create community apart from heteroreality and yet still engage the world socially and politically with the intent to make a difference in the world.[16] My Whiteheadian interpretation of worldly separatism is that women are self-creating, valuable beings with freedom to select from a world of possibilities and relationships how they will form themselves. To understand myself as self-creating and free means that I must likewise honor and respond to the diversity of women who may choose different options and relationships. As an activist Whiteheadian, who accepts that relationships are central to our deepest spiritual and personal identity, I propose that our activism must include engagement of a broad range of relationships across lines of gender, race, class, religion, and culture. An engaged worldly separatism permits personal, social, and political creativity by immersion in diverse relationships that entail facing the complexity of difference and diversity among humans and nonhuman nature.

---

[14] Ibid., 191–92.

[15] Ibid., 188–89.

[16] Janice G. Raymond, *A Passion for Friends: Toward a Philosophy of Female Affection* (Boston: Beacon, 1986) 144, 153.

Proposal 3:
Dualism and Hierarchy

A fundamental problem for feminism is how to interpret difference, especially gender difference, but other kinds of differences are also at stake for ecofeminism. The literature influencing *A Feminist Cosmology* prepared the groundwork for discussion of dualism and hierarchy. Often feminist literature argues that dualism and hierarchy are oppressive interpretations of differences. The book reminds readers that Patricia Hill Collins and others understand that dichotomizing thinking, which tends to value masculine over feminine and human over nature, is the foundation for social hierarchy.[17] Karen Warren, who is also cited, notes that hierarchy coupled with a "logic of domination" creates an oppressive system even if hierarchy itself need not be a problematic concept.[18] Regardless of the empirical or ontological significance of dualism and hierarchy, value assessments of differences have axiological consequences.

Feminist (and ecofeminist) deconstruction of hierarchy and dualism is a notable philosophical and theological accomplishment of the last four decades. With the presumption that social justice and ecojustice go hand in hand, ecofeminists have tried to figure exactly what so galls us about hierarchy, dualism, and domination. Ecofeminist discomfort with hierarchy and dualism, for example in Karen Warren's philosophy, is not so much concerned about defining differences using dichotomized pairs as with the foundational use of dualism to create hierarchical systems of oppression and domination. Some ecofeminists recognize that hierarchy is a characteristic of social and biological relationships in nature. Others argue a decidedly antihierarchical and nondualistic interpretation of nature.

My question to myself and ecofeminist theologians generally is, How can we refine our criticism of hierarchy and dualism? We are quite justified in our complaint that hierarchical thinking is forgetful of context and criteria. Abstract hierarchical thinking loses track of its biology-based essentialism. At the same time, it forgets that rationality and cognitive complexity might not be the only criteria by which we might compare and evaluate the order of nature, including ourselves.

While we ecofeminists may be right about some things, ecofeminist theology is equally as abstract as hierarchical, patriarchal theology when

---

[17] Patricia Hill Collins, *Black Feminist Thought: Knowledge, Consciousness, and thee Politics of Enpowerment* (New York: Routledge, 1990) 70.

[18] Karen J. Warren, "The Power and the Promise of Ecological Feminism," *Environmental Ethics* 12.2 (1990) 128.

we write about nature. Our ideology may be noble, even just, but we have derived our theology from a rather generalized engagement of science, using a few very general themes from ecological science. Rarely have ecofeminist theologians looked at nature—*really* looked at nature. Mostly we look at humans and what humans do to nature. Because ecofeminism does not start with nature empirically and reflect on humans as a species in nature, we may be missing critical sources for ecofeminist praxis. Our ethics are nicely supported, but our view of nature is still too romantic and humanocentric.

One specific puzzle in my work, for example, arises from conversation with biologists and social scientists who study primates. Scientists use the language of *dominance hierarchy* to indicate that aggression and rank are characteristics of animal behavior that order social relations and contribute to survival. My recent research project studying chimpanzees undeniably convinces me that aggression and violence are common patterns. As an ecofeminist, I am as disappointed as Jane Goodall was to learn that chimpanzees are very violent hunters of monkeys, especially since chimpanzees have richer nutritional options. I am troubled by field observation of lethal struggles for alpha male position, brutal and forced mating with females, and systematic massacre of adjacent chimpanzee groups. Arguably, science is a social construction to some extent, yet observations of aggression are recognized by feminist scientists who bring ideologies similar to those of feminist theologians to primate field observations. However, even with clear and replicated observations of aggression, dominance hierarchy is not elevated necessarily to an abstract philosophical or ethical principle.

One further question for ecofeminism, then, is, In light of less generalized reference to science, what nuances could ecofeminists bring to the criticism of hierarchy, dualism, and dominance? Ecofeminists can still refute the logic of domination, and we can advocate the moderation of aggression among humans, but what shall we say about dominance in nature if we *are* nature? Once ecofeminists grant the presence of domination, aggression, rank, and hierarchical order, we cannot simply concede the simplistic argument that "hierarchy exists in nature and, therefore, hierarchy among humans and over nature is justified." Simply asserting an antihierarchical stance is not as effective as examining the nuances of dominance hierarchy in nature and understanding the empirical evidence that helps advance an argument against oppression. For example, we need not accept that nature prescribes corporate hierarchies as such or abuse, subordination, and exploitation of women and nature. Paul Ehrlich summarizes some of the scientific consensus important for understanding dominance hierarchy and reverse dominance hierarchy among humans:

> Thus, in the transition to modern hunting and gathering *Homo sapiens,* there must have been a general trend toward softening chimp-like dominance hierarchies, enabling evolution of the increasingly egalitarian, nonstratified societies that many scholars believe were characteristic of our hunter-gatherer ancestors. In them coalitions presumably would have limited the power of otherwise dominant individuals.[19]

Later in the same passage, Ehrlich writes:

> Such social sanctions, ranging from ridicule to assassination, against disruptive individuals are widely reported in hunter-gatherer societies. This supports the hypothesized trend toward more egalitarian societies within these groups and a loss of overwhelming control by dominant males—people cooperate to dispose of those who possess what is judged to be too much power.[20]

Ehrlich's interpretation of the scientific evidence assists us in examining the development of less stratified and hierarchical social organizations of humans—information helpful in defending a more egalitarian understanding of male and female humans without requiring gender essentialism or gender identity. Similarly, such information would be helpful in discussion of male-female dualism or polarity, which (I aver) must be subject to empirical evidence from the biological and social sciences in considering broad population and individual characteristics and differentiations.

While feminism tends to focus on gender relationships—male and female, as well as female-female—ecofeminism also addresses the problem of dualism and hierarchy in nonhuman nature. The issues are often addressed in terms of the nature-culture dualism, because exploitation of nature is often justified by distancing humans from nature and asserting that culture is more valuable than nature. The discussion of hierarchy, dualism, and dominance leads easily to reflection on nature and culture. To what extent are hierarchy and dominance natural phenomena or cultural artifacts? More broadly, what refinements might ecofeminists make in critical and constructive thinking about nature and culture?

---

[19] Paul R. Ehrlich, *Human Natures: Genes, Cultures, and the Human Prospect* (Washington, D.C.: Island, 2000) 209.

[20] Ibid.

Chimpanzees again are interesting resources for rethinking ecofeminist approaches to nature. In 1999, Andrew Whiten, Jane Goodall, and other primatologists collaboratively published an article in the journal *Nature*.[21] The article compiled and compared data from decades of observations of seven African groups of chimpanzees. Comparisons of data suggested that some regularly observed behaviors could not be attributed to either genetics or environment. The behaviors were (and are) most likely transmitted by learning events, and the scientists called their findings the discovery of chimpanzee culture. Admittedly, the definition of *culture* satisfies biologists and not social scientists, but at least some researchers give theologians data and interpretations of culture in animals we have associated solely with nature (except in film and children's books) to consider in reflections. If so-called nature exhibits culture, how might ecofeminist theology refine our discussions of the relationship of nature and culture?

Nancy Frankenberry's criticism of the naturalization of women pushes ecofeminists to refine reflection on nature and culture in other ways.[22] *The Less Noble Sex* by Nancy Tuana surveys the historical relationship of science, philosophy, and religion and their common tendency to argue that women and Africans are not-quite-human or less evolved than European males.[23] Designating groups of humans as more like nature or like animals justified diverse practices from withholding education to binding into slavery. In response, some ecofeminists attempted to reinvent or repristinate the naturalization of women and the oppressed, but perhaps were not so astute about the subtleties in associating humans with nature. For example, men are commonly associated with nature, too—for athletic performance, sexual prowess, or boorish behavior. What kinds of details about naturalization of humans are missing from ecofeminist crossing of the nature-culture divide? What is the next stage of refinement in dismantling nature-culture dualism?

The third proposal for ecofeminist and feminist theology is that we must refine our reflections on hierarchy and dualism using empirical evidence from science to provide nuance. To continue the progress of my own scholarship,

---

[21] Andrew Whiten, Jane Goodall, W. C. McGrew, T. Nishida, V. Reynolds, Y. Sugiyama, C. E. G. Tutin, R. W. Wrangham, and C. Boesch, "Cultures in Chimpanzees," *Nature* 399 (1999) 682–85.

[22] Nancy Frankenberry, *Religion and Radical Empiricism*, SUNY Series in Religious Studies (Albany: State University of New York Press, 1987).

[23] Nancy Tuana, *The Less Noble Sex: Scientific, Religious, and Philosophical Conceptions of Woman's Nature*, Race, Gender, and Science (Bloomington: University of Indiana Press, 1993).

I propose that I need to explore in more detail how dominance hierarchy is defined and how the science-religion dialogue, feminist, and gender biased approaches may actually equivocate by using the same terms to refer to different phenomena or ideas. As part of the struggle to write more adequately about dualism and hierarchy, my work needs to engage in deconstruction of whiteness and other structures of oppression.

In conclusion then, I find my own ecofeminist project to be unfinished, and my proposals are a public acknowledgment of emerging projects. Three projects are high priorities:

- Exploring the promise of attention epistemology and narrative as methodological approaches to both whole and parts in nature in order to balance generalizations bout the ecosystem with insight from the depths of species and individual lives of animals or other creatures in nature.

- Seeking coalitions that give insight and perspective and that support the justice that I advocate.

- Refining reflections on hierarchy and dualism using empirical evidence from science to provide nuance.

In moving beyond *A Feminist Cosmology*, I hope that I listen well to those who engage and criticize my scholarship. I value their collaboration with me and join their critique with an eye toward new projects.

Bibliography

Abe, Masao. *Zen and Western Thought.* Edited by William R. LaFleur. Honolulu: University of Hawaii Press, 1985.

Abram, David. *The Spell of the Sensuous: Perception and Language in a More-than-Human World.* New York: Pantheon, 1996.

Adams, Carol J. *The Sexual Politics of Meat: A Feminist-Vegetarian Critical Theory.* New York: Continuum, 1990.

Allen, Prudence. *The Concept of Woman: The Aristotelian Revolution, 750 BC—AD 1250.* 1985. Reprinted, Grand Rapids: Eerdmans, 1997.

Arendt, Hannah. *The Human Condition.* Chicago: University of Chicago Press, 1958.

Beauvoir, Simone de. *Deuxième sexe.* Paris: Gallimard, 1949.

———. *The Second Sex.* Translated by H. M. Parshley. New York: Knopf, 1953.

Birch, Charles, and John B. Cobb Jr. *The Liberation of Life: From Cell to the Community.* Cambridge: Cambridge University Press, 1981.

Butler, Judith. *Gender Trouble: Feminism and the Subversion of Identity.* New York: Routledge, 1990.

Chodron, Bhikshuni Thubten. *Interfaith Insights.* New Delhi: Timeless, 2000.

Crites, Stephen, "The Narrative Quality of Experience." In *Why Narrative? Readings in Narrative Theology,* edited by Stanley Hauerwas and L. Gregory Jones. 1989. Reprinted, Eugene, Ore.: Wipf & Stock, 1997.

Daley, Mary. *Beyond God the Father: Toward a Philosophy of Women's Liberation.* Boston: Beacon, 1973, 1985.

Darlington, Susan M. "Tree Ordination in Thailand." In *Dharma Rain: Sources of Buddhist Environmentalism,* edited by Stephanie Kaza and Kenneth Kraft, 198–205. Boston: Shambhala, 2000.

De Waal, Frans. *The Ape and the Sushi Master: Cultural Reflections of a Primatologist.* New York: Basic, 2001.

Dogen. "Genjo Koan: Actualizing the Fundamental Point." In *Moon in a Dewdrop,* edited by Kazuki Tanahashi. San Francisco: North Point Press, 1985.

Dumoulin, Heinrich. *Zen Buddhism: A History.* Vol. 1: *India and China.* Translated by James W. Heisig and Paul Knitter. New York: Macmillan, 1988.

Ehrlich, Paul R. *Human Natures: Genes, Cultures, and the Human Prospect.* Washington, D.C.: Island, 2000.

Fox, Matthew. *Illuminations of Hildegard of Bingen.* Santa Fe, N.M.: Bear, 1985.

Fox, Warwick. *Toward a Transpersonal Ecology: Developing New Foundations for Environmentalism.* Boston: Shambhala, 1990.

Frankenberry, Nancy. "The Earth Is Not Our Mother: Ecological Responsibility and Feminist Theory." In *Religious Experience and Ecological Responsibility*, edited by Donald A. Crosby and Charley D. Hardwick, 23–50. New York: Lang, 1996.

Friedan, Betty. *The Feminine Mystique*. New York: Norton, 1963.

Gilman, Charlotte Perkins. *Moving the Mountain*. New York: Charlton, 1911.

Goodall, Jane. Forward to *Minding Animals: Awareness, Emotions, and Heart*, by Mark Bekoff. Oxford: Oxford University Press, 2002.

———. *Through a Window: My Thirty Years with the Chimpanzees of Gombe*. Boston: Houghton Mifflin, 1990.

———, and Philip Berman. *Reason for Hope: A Spiritual Journey*. New York: Warner, 1999.

Grene, Marjorie. *Dreadful Freedom: A Critique of Existentialism*. Chicago: University of Chicago Press, 1948.

Grosnick, William. "The Buddhahood of the Grasses and the Trees." In *An Ecology of Spirit*, edited by Michael Barnes, 197–208. 1990. Reprinted, Lanham, Md.: University Press of America, 1994.

Gross, Rita M. *Buddhism after Patriarchy: A Feminist History, Analysis, and Reconstruction of Buddhism*. Albany: State University of New York Press, 1993.

Guenther, Herbert V., and Leslie S. Kawamura, translators. *Mind in Buddhist Psychology*. Tibetan Translation Series. Emeryville, Calif.: Dharma, 1975.

Hanh, Thich Nhat. "The Sun in My Heart." In *Love and Action: Writings on Nonviolent Social Change*. Berkeley: Parallax, 1993.

Harris, Ian. "Buddhism and Ecology," In *Contemporary Buddhist Ethics*, edited by Damien Keown, 113–35. Surrey, Eng.: Curzon, 2000.

Hauerwas, Stanley. "Story and Theology." In *Truthfulness and Tragedy: Further Investigations in Christian Ethics*, 71–81. Notre Dame: University of Notre Dame Press, 1977.

Heyward, Carter. *The Redemption of God: A Theology of Mutual Relation*. Washington, D.C.: University Press of America, 1982.

Hill Collins, Patricia. *Black Feminist Thought: Knowledge, Consciousness, and the Politics of Empowerment*. Perspectives on Gender 2. Boston: Unwin Hyman, 1990.

Honghzi. *Cultivating the Empty Field: The Silent Illumination of Zen Master Hongzhi*. Translated by Taigen Daniel Leighton with Yi Wu. San Francisco: North Point, 1991.

Howell, Nancy R. *A Feminist Cosmology: Ecology, Solidarity, and Metaphysics*. Amherst, N.Y.: Humanity, 2000.

Jantzen, Grace. *Becoming Divine: Toward a Feminist Philosophy of Religion*. Bloomington: Indiana University Press, 1999.

Kalupahana, David J. *The Principles of Buddhist Psychology*. SUNY Series in Buddhist Studies. Albany: State University of New York Press, 1987.

Kaza, Stephanie. "Acting With Compassion: Buddhism, Feminism, and the Environmental Crisis." In *Ecofeminism and the Sacred*, edited by Carol Adams, 50–69. New York: Continuum, 1993.

———, and Kenneth Kraft, editors. *Dharma Rain: Sources of Buddhist Environmentalism*. Boston: Shambhala, 2000.

Keller, Catherine. *From a Broken Web: Separation, Sexism and Self*. Boston: Beacon, 1986.

LaFleur, William R. "Sattva: Enlightenment for Plants and Trees." In *Dharma Gaia: A Harvest of Essays in Buddhism and Ecology*, edited by Alan Hunt Badiner, 136–44. Berkeley: Parallax, 1990.

Macy, Joanna. "Faith, Power and Ecology." In *World as Lover, World as Self*. Berkeley: Parallax, 1991.

Marcel, Gabriel. *Being and Having*. Translated by Katherine Farrer. 1949. Reprinted, New York: Harper & Row, 1967.

————. *Creative Fidelity*. Translated by Robert Rosthal. New York: Farrar, Straus, 1964.

————. *The Mystery of Being*. Vol. 1. Translated by G. S. Fraser. Chicago: Regnery, 1950.

Marcos, Sylvia. "Cognitive Structures and Medicine: The Challenge of Mexican Popular Medicines." *Curare* 11.2 (1988) 87–96.

————. "Gender and Moral Precepts in Ancient Mexico: Sahagun's Texts." *Concilium*, edited and translated by Jacqueline Mosio 6 (1991) 60–74.

————. "Embodied Religious Thought: Gender Categories in Mesoamerica." *Religion* 28 (1998) 371–82.

McFague, Sallie. *Metaphorical Theology: Models of God in Religious Thought*. Philadelphia: Fortress, 1982.

————. *The Body of God: An Ecological Theology*. Minneapolis: Fortress, 1993.

————. *Super, Natural Christians: How We Should Love Nature*. Minneapolis: Fortress, 1997.

————. *Life Abundant: Rethinking Theology and Economy for a Planet in Peril*. Minneapolis: Fortress, 2001.

McTernan, Vaughan. "Performing God: God, the Organic and Postmodernism." *American Journal of Theology and Philosophy* 23 (2002) 236–51.

Newman, Barbara. *Sister of Wisdom: St. Hildegard's Theology of the Feminine*. Berkeley: University of California Press, 1987.

Plumwood, Val. *Feminism and the Mystery of Nature*. New York: Routledge, 1993.

Primavesi, Anna. *Sacred Gaia: Holistic Theology and Earth System Science*. London: Routledge, 2000.

Queen, Christopher, editor. *Engaged Buddhism in the West*. Somerville, Mass.: Wisdom, 2000.

Raymond, Janice G. *A Passion for Friends: Toward a Philosophy of Female Affection*. Boston: Beacon, 1986.

Riceour, Paul. *Interpretation and Theory: Discourse and the Surplus of Meaning*. Fort Worth: Texas Christian University Press, 1976.

————. *Time and Narrative*. 3 vols. Translated by Kathleen McLaughlin and David Pellauer. Chicago: University of Chicago Press, 1984–88.

————. *Oneself as Another*. Translated by Kathleen Blamey. Chicago: University of Chicago Press, 1992.

Rolston, Holmes III. "Values in Nature." In *Philosophy Gone Wild*. Buffalo, N.Y.: Promethus, 1989.

Ruether, Rosemary Radford, "Misogynism and Virginal Feminism in the Fathers of the Church." In *Religion and Sexism: Images of Women in the Jewish and Christian Traditions*, edited by Rosemary Radford Ruether, 150–83. New York: Simon and Schuster, 1974.

————. *Gaia and God: An Ecofeminist Theology of Earth Healing*. San Francisco: HarperSanFrancisco, 1992.

Salzberg, Sharon. *Loving Kindness: The Revolutionary Art of Happiness*. Boston: Shambala, 1965.

Sartre, Jean-Paul. *Being and Nothingness: An Essay on Phenomenological Ontology*. Translated by Hazel E. Barnes. London: Methuen, 1957.

Schrag, Calvin O. *The Self after Postmodernity*. New Haven: Yale University Press, 1997.

Sivaraksa, Sulak. *Seeds of Peace: A Buddhist Vision for Renewing Society*. Edited by Tom Ginsburg. Berkeley: Parallax, 1992.

———. *Santi Pracha Dhamma*. Bangkok: Santi Pracha Dhamma Institute, 2001.

Snyder, Gary. *The Practice of the Wild: Essays*. San Francisco: North Point, 1990.

———. *A Place in Space: Ethics, Aesthetics, and Watersheds—New and Selected Prose*. Washington, D.C.: Counterpoint, 1995.

Suchocki, Marjorie. *The End of Evil: Process Eschatology in Historical Context*. 1988. Reprinted, Eugene, Ore.: Wipf & Stock, 2005.

Swimme, Brian. *The Universe Is a Green Dragon: A Cosmic Creation Story*. Santa Fe, N.M.: Bear,1984.

———, and Thomas Berry. *The Universe Story: From the Primordial Flaring Forth to the Ecozoic Era—A Celebration of the Unfolding of the Cosmos*. San Francisco: HarperSanFrancisco, 1992.

Szerszynski, Bronislaw. "The Varieties of Ecological Piety." *Worldviews: Environment, Culture, Religion* 1 (1997) 37–56.

Tuana, Nancy. *The Less Noble Sex: Scientific, Religious, and Philosophical Conceptions of Woman's Nature*. Race, Gender, and Science. Bloomington: University of Indiana Press, 1993.

Van Ewijk, Thomas J. M. *Gabriel Marcel: An Introduction*. Translated by Matthew J. van Velzen. Glen Rock, N.J.: Paulist, 1965.

Walker, Alice. *The Color Purple*. New York: Washington Square, 1982.

Walker, Susan, editor. *Speaking of Silence*. New York: Paulist, 1987.

Warren, Karen J. "The Power and Promise of Ecological Feminism." *Environmental Ethics* 2 (1990) 132–46.

———. *Ecofeminist Philosophy: A Western Perspective on What It Is and Why It Matters*. Studies in Social, Political, and Legal Philosophy. Lanham, Md.: Roman and Littlefield, 2000.

Whitehead, Alfred North. *Process and Reality*. 1929. Corrected ed. Edited by David Ray Griffin and Donald W. Sherburne. New York: Free Press, 1978.

———. *Adventures of Ideas*. 1933. New York: Free Press, 1961.

———. *Modes of Thought*. 1938. Reprinted, New York: Free Press, 1968.

Whiten, Andres, et al. "Cultures in Chimpanzees." *Nature* 399 (1999) 682–85.

Wilson, Marie. "Wings of the Eagle." In *Healing the Wounds: The Promise of Ecofeminism*, edited by Judith Plant, 212–18. Philadelphia: New Society, 1989.

Young, Serinity, editor. *An Anthology of Sacred Texts by and about Women*. New York: Crossroad, 1993.

Contributors

Kathlyn A. Breazeale is Assistant Professor of Religion in the area of Contemporary Theology, with a specialty in Feminist and Womanist Theologies, at Pacific Lutheran University. She is a graduate of the Claremont Graduate School, where her focus was on feminist and process theologies. For stimulating her thinking regarding the constructive aspect of her article in this volume, She would like to express appreciation to the students in her course "Women, Nature and the Sacred," both at Pacific Lutheran University and Prescott College. She is the author of several articles dealing with issues of sexuality from a feminist process theological perspective, including "Don't Blame It on the Seeds," "Marriage after Patriarchy?" and the forthcoming "Process Perspectives on Love, Sexuality and Marriage." Other areas of research and publication include the intersection of feminist theology, the arts, and social justice; religion and public life; and feminist pedagogy. Her current projects include a theological and pedagogical analysis of the social activist art of Corita Kent and a book manuscript, *Partners after Patriarchy: Toward a Theology of Redemptive Intimacy*. Kathi also expresses her theological and pedagogical interests in her work as a liturgical dancer.

Nancy R. Howell is Professor of Theology and Philosophy of Religion at Saint Paul School of Theology, a United Methodist seminary in Kansas City, Missouri. She is author of *A Feminist Cosmology: Ecology, Solidarity, and Metaphysics* and associate editor of the *Encyclopedia of Science and Religion*. Howell serves on the editorial boards of the *Journal of the American Academy of Religion,* the *Journal of Religion and Abuse,* and the *American Journal of Theology and Philosophy.* She serves on the boards of the Center for Process Studies, the Highlands Institute for American Religious and Philosophical Thought, and the Metanexus Institute on Religion and Science.

Paul O. Ingram is Professor of Religion at Pacific Lutheran University, Tacoma, Washington, where he teaches history of religions. He is the author of five books, including *The Modern Buddhist-Christian Dialogue,*

Wrestling With the Ox: A Theology of Religious Experience, and editor (with Sallie B. King, James Madison University) of *The Sound of Liberating Truth: Buddhist-Christian Dialogues in Honor of Frederick J. Streng*, which received the Society for Buddhist-Christian Dialogue's Frederick J. Streng Book of the Year Award for 2003. He served as President of the Society for Buddhist-Christian Studies (1998-2000) and currently serves on its Executive Committee.

Stephanie Kaza is Associate Professor of Environmental Studies at the University of Vermont, where she teaches religion and ecology, Buddhism and ecology, ecofeminism, and unlearning consumerism. Dr. Kaza is vice-president of the Society for Buddhist-Christian Studies and a member of the International Christian-Buddhist Theological Encounter group. She is the author of *The Attentive Heart: Conversations with Trees*, and co-editor (with Kenneth Kraft) of *Dharma Rain: Sources of Buddhist Environmentalism*, which is an anthology of classic and modern texts supporting a Buddhist approach to environmental activism. Her newest work is the edited volume, *Hooked! Buddhist Writings on Greed, Desire, and the Urge to Consume.*

Lisa Stenmark earned her MA in Systematic Theology from the Graduate Theological Union and her PhD in Religious Studies from Vanderbilt University. She is the founder and Director of Women in Religion, Ethics and the Sciences (WiRES) and the Associate Director of the Institute for Social Responsibility, Ethics and Education at San Jose State University, where she teaches in the Comparative Religious Studies Program. She has been active in the science and religion discourse for almost a decade, winning the Templeton Course Prize in 1998, teaching, presenting and publishing papers, and serving as the co-chair of the American Academy of Religion's Science and Religion Group. Her scholarly interests include the implications of narrative trajectories for understanding the relationship between science, technology, and religion and rethinking the ways that religion, science, and the science and religion discourse can and should engage in the public sphere. Her current project is a collaboration with William Stahl entitled *Deep Narrative: Myth, Meaning and Modernity*. In her spare time she trains for triathlons and is still an avid trekker.

Marjorie Hewitt Suchocki is Professor Emerita from Claremont School of Theology, where she held the Ingraham Chair in theology. She has written many articles, chapters in books, and books—most recently *Divinity & Diversity: A Christian Affirmation of Religious Pluralism*. She currently directs the Whitehead International Film Festival and the Process & Faith Program of the Center for Process Studies at Claremont.

Marit A. Trelstad is Assistant Professor of Religion at Pacific Lutheran University, Tacoma, Washington, where she teaches constructive theology. Her dissertation was entitled *Defining the Self in a Relational Philosophical Theology*. Subsequently, she has published articles and presented papers connecting her interest in theological anthropology and the philosophy of religion to issues of theology and violence to women. Her current writing focuses on the theology of the cross, the possession of women, and pastoral counseling with rape survivors. She is currently a fellow with the Wabash Center for Teaching and Learning in Theology and Religion.

www.ingramcontent.com/pod-product-compliance
Lightning Source LLC
Chambersburg PA
CBHW070926270326
41927CB00011B/2740